INTERNATIONAL SERIES OF MONOGRAPHS IN
ANALYTICAL CHEMISTRY
GENERAL EDITORS: R. BELCHER AND H. FRIESER

VOLUME 39

THE DESTRUCTION OF ORGANIC MATTER

THE DESTRUCTION OF ORGANIC MATTER

BY

T. T. GORSUCH

B.Sc., Ph.D., A.R.I.C.

Chief Scientific Executive, RHM Foods Ltd.

PERGAMON PRESS

Oxford · New York · Toronto · Sydney · Braunschweig

Pergamon Press Ltd., Headington Hill Hall, Oxford
Pergamon Press Inc., Maxwell House, Fairview Park, Elmsford, New York 10523
Pergamon of Canada Ltd., 207 Queen's Quay West, Toronto 1
Pergamon Press (Aust.) Pty. Ltd., 19a Boundary Street, Rushcutters Bay,
N.S.W. 2011, Australia
Vieweg & Sohn GmbH, Burgplatz 1, Braunschweig

First edition 1970

Library of Congress Catalog Card No. 71-109583

PRINTED IN GREAT BRITAIN BY A. WHEATON & CO., EXETER

08 015575 8

CONTENTS

PREFACE   vii

CHAPTER 1. General Introduction   1

CHAPTER 2. Methods Not Involving Complete Destruction   3

CHAPTER 3. Sources of Error   9

CHAPTER 4. Wet Oxidation   19
    A.   Sulphuric acid   19
    B.   Nitric acid   21
    C.   Perchloric acid   22
    D.   Hydrogen peroxide   24
    E.   Mechanism   24
    F.   Problems   26

CHAPTER 5. Dry Oxidation   28
    A.   General discussion   28
    B.   Methods of dry ashing   31
       1. Closed systems   31
       2. Open systems   33
    C.   Oxidation with excited oxygen   39

CHAPTER 6. Oxidative Fusion and Other Methods   40
    A.   Oxidative fusion   40
    B.   Oxidation in an atmosphere of nitric acid 42
    C.   Oxidation with ozone   42

CHAPTER 7. Methods of Investigation   43
    A.   General discussion   43
    B.   Choice of a radioactive tracer   48

CHAPTER 8. Individual Elements 55
- A. Lithium, sodium, potassium, rubidium and caesium 55
- B. Copper, silver and gold 60
- C. Beryllium, magnesium, calcium, strontium, barium and radium 67
- D. Zinc, cadmium and mercury 73
- E. Aluminium, gallium, indium, thallium 84
- F. Scandium, yttrium and the rare earths 87
- G. The actinide elements 88
- H. Germanium, tin and lead 89
- I. Titanium, zirconium and hafnium 99
- J. Arsenic, antimony and bismuth 100
- K. Vanadium, niobium and tantalum 112
- L. Chromium, molybdenum and tungsten 114
- M. Manganese, technetium and rhenium 116
- N. Selenium, tellurium and polonium 117
- O. Iron, cobalt and nickel 121
- P. The platinum metals 134

CHAPTER 9. Selected Decomposition Procedures 136
- A. Wet digestion methods 137
  - 1. Nitric and sulphuric acids 138
  - 2. Mixtures containing perchloric acid 141
  - 3. Sulphuric acid and hydrogen peroxide 142
- B. Dry ashing procedures 143
  - Ashing aids 144

APPENDIX 1. Nuclear Data on Radioactive Tracers 145

INDEX 149

OTHER TITLES IN THE SERIES IN ANALYTICAL CHEMISTRY 152

PREFACE

THIS monograph has its origins in work carried out in 1956, 1957 and 1958 during the author's tenure of the first research fellowship awarded by the Society for Analytical Chemistry. During the course of this work investigating the losses that can occur during wet and dry oxidation, it became clear that this step was probably the greatest single source of error in the majority of trace element determinations, yet in many publications it rated the very briefest of mentions: "The sample was ignited in a porcelain basin", or even "The material was ashed . . .".

Since those years much more investigational work has been published on this topic, and the moment now seems opportune to try to collect some of it together.

The author's own major field of interest in determinations involving traces of elements, and his bias towards radiochemical methods of investigation will be abundantly clear to anyone who opens this book, but for this there seems no need for apology. The trace level determinations are usually the more demanding, and the radiochemical methods seem by far the most rewarding.

Some explanation is perhaps required for a certain amount of repetition that is to be found. This is intentional, to allow separate sections of the book to be read alone, without the need to search the other chapters for examples. Many people will be mainly interested in specific elements or groups of elements, or be primarily concerned with mechanisms, or reagents, or methods. Each of these it is hoped, can with profit read a part of the book, independent, in the first instance, of the others.

The number of references in the literature to methods of decomposition of organic samples is immense. To attempt to quote them all, or even most of them would be impossible and indeed, pointless. The references quoted are sometimes ones picked merely as examples of a particular method or mechanism but in the main an attempt has been made to include the more significant publications in the field. Inevitably many will have been missed as relevant publications occur in hundreds of journals covering the whole

vii

spectrum both of science and of craft, and I would be grateful to have important contributions to the subject brought to my notice.

I gratefully acknowledge the assistance of my wife and of my secretary, Miss Teresa Butterworth, in preparing this manuscript, but claim all credit for the sins of omission and commission it contains. I have enjoyed writing it: I hope it will be of some small value.

T. T. GORSUCH

CHAPTER 1

GENERAL INTRODUCTION

ONE OF the commonest measurements called for in analysis is the determination of an element present in an organic compound or an organic
matrix. Such an analysis usually involves several operations, and this book
is concerned with just a part of one of these stages.

The main structural components of organic compounds—carbon,
hydrogen, oxygen, nitrogen, sulphur, phosphorus, silicon and the halogens
—are excluded from consideration, as are boron and the rare gases, but all
the remaining metals and metalloids are considered to fall within its scope.

The problems involved in dealing with such a heterogeneous group are
increased by the fact that very many of the elements to be considered can
occur in organic materials in one, or both, of two ways. Firstly metals
and metalloids can occur in chemical combination with the "organic"
elements listed above, to give clearly defined chemical compounds. In
some instances the form of combination is such that the "inorganic"
element retains many of its normally assumed properties, as with the alkali
metal salts of simple organic acids, but in other cases the organometallic
compounds formed largely or entirely suppress the properties associated
with the element in its inorganic form.

The second mode of occurrence of these "other elements" in organic
matrices is as minor components which bear no clearly defined relationship
to the bulk of the sample. Often their occurrence is as a result of chance
contamination, perhaps during the storage of a food, or the processing of
a bulk chemical, but in many situations the element itself exists as a
distinct organic compound which is distributed through a large amount of
other organic material. An example of this may be found in mammalian
tissues where part of the iron present is likely to be combined in haemo-
globin distributed throughout the tissue.

These two categories, of course, shade into one another, from the pure
organometallic compound, through the slightly impure sample, right down

to the organic matrix with only a few parts per million, or parts per billion of a foreign element present in it.

The basic form of virtually any determination is that the object of the determination, or some suitable property of it, is separated from all sources of interference and then measured. In many of the element determinations with which we are concerned the organic matter present in the sample is the principal source of interference, and a common prelude to the measurement process is a procedure to eliminate or reduce this interference, generally by destroying or disrupting the organic material.

The extent to which this disruption of the sample must be pursued, before interference in the final determination is minimized or eliminated, varies widely. In some instances no treatment is required at all; in others all the organic material must be destroyed and between these extremes come all the possible intermediate degrees of decomposition.

This monograph deals mainly with the total decomposition procedures, but some brief consideration is also given to techniques which fall short of this.

CHAPTER 2

METHODS NOT INVOLVING COMPLETE DESTRUCTION

ALTHOUGH the major part of this monograph is concerend with methods of sample preparation which involve the complete destruction of the organic matrix, there are a considerable number of procedures which require substantially less than this to make the metal content of the sample available for determination. Indeed, in a number of instances it is possible to complete the measurements without any kind of chemical attack upon the sample.

Activation analysis can, under favourable conditions, be virtually non-destructive. Many elements, when bombarded with slow neutrons, give rise to radioactive species, and these often emit gamma-radiation. When the other components of the sample do not interfere, it is possible to identify and determine the radioactive species by means of the gamma photons that they emit, and from this, to measure the amount of the element present in the sample. An easy example of the application of this technique can be found in the determination of sodium in fuel oil where the very energetic gamma of $2\cdot75$ MeV is readily measured. The biggest drawback to this method is, perhaps, the cost involved: access to a source of neutrons is necessary—for good sensitivity this needs to be a nuclear reactor—and a considerable amount of electronic equipment is required.

Another method which utilizes the emission of radiation from a sample as a result of bombardment, is X-ray fluorescence analysis. In this instance the bombarding radiation is X- or gamma-rays from a suitable source, while the emitted radiation is the characteristic X-rays which arise from the components of the sample. This technique has found considerable application in the oil industry, but there have been numerous other uses also such as, the determination of catalysts in low-pressure polythene, of silver in photographic materials, and of various elements in plant and animal materials. The method is extremely rapid, but in complex materials is likely to suffer from interference. It is best used when one or a few elements are to be determined in an essentially constant matrix and where there are many similar samples to measure so that the necessary, and generally lengthy, calibrations and corrections are worth carrying out.

A further non-destructive method, involving radiation, depends upon the extent to which electromagnetic radiation is absorbed by elements of different atomic number. In general the absorption coefficient for low energy X- or gamma-rays (less than 100 keV) rises steeply with increasing atomic number, so that a high atomic number material occurring in a sensibly constant matrix of lower atomic number can be determined by measuring the degree to which radiation of a suitable energy is absorbed during passage through the sample. As with other absorptiometric methods the expression relating to the intensity of the incident radiation (I_0) and the intensity of the transmitted radiation (I) is in the form:

$$I = I_0 e^{-\mu \rho t}$$

where μ = the mass absorption coefficient, ρ = the density and t the thickness. The product ρt is equal to the mass of the absorber per unit area and variations in absorption can be used to measure this. When the sample consists of a solution of a heavy element in a matrix of low atomic number, then by choosing a radiation of suitable wavelength, variations in the concentration of the heavy element can be determined. This type of measurement has been applied in a number of specialized situations such as the determination of lead in petroleum as it is again very rapid and lends itself well to continuous monitoring of process streams. A related technique which has been used for similar determinations is the back-scatter measurement method. In this method, instead of measuring the radiation absorbed by the sample in transmission, the important parameter is the amount of radiation reflected or scattered. If thick samples and beta-radiation are used, the ratio of the intensity of the scattered radiation to the intensity of the incident radiation varies with the atomic number (Z) of the sample, being about $0 \cdot 16$ for $Z = 10$ and about $0 \cdot 80$ for $Z = 80$. These methods can therefore again be used for determining heavy elements in a matrix of low atomic number.

A non-destructive technique of rather limited application is electron spin resonance, which can be used to detect and determine small amounts of elements or ions which have one or more unpaired electrons. Many elements of technical importance do fall into this group, and a typical application of the method is the determination of vanadium in petroleum products.

Another limited technique, though a much less expensive one than electron spin resonance, is the use of certain cation-sensitive glass electrodes. This technique has been used for the determination of the sodium content of bacon merely by inserting the electrode into a hole cut in a rolled slice of bacon.

Several other procedures have been used for the determination of various elements in organic matrices, which, although they do not require any previous treatment of the sample, do alter it in some way in the determination itself. In the main these methods can be used either where the element occurs as some specific compound which can be determined, or where it occurs in a matrix which does not interfere in the determination.

Mass spectrometry and gas chromatography are two techniques by which specific compounds of lead have been identified, and determined, in petrol samples, and both are fields capable of extension. Gas chromatography of metal complexes particularly has expanded considerably with the use of heat-stable complexing agents, but its application to the determination of metals in organic materials has so far been limited.

Many titrimetric procedures have been described, both for specific organo-metallic compounds, and for elements present in ionic form in organic matrices. Methods used for organo-metallic compounds range from the oxidative titration of some organic arsenicals with iodine to the determination of alkyl zinc and alkyl aluminium compounds by thermometric titration with organic titrants. Essentially ionic elements in organic mixtures have mainly been determined by complexometric titration. Organic samples such as lubricating oils or paint dryers can be dissolved in organic solvent mixtures and the metallic components titrated with EDTA, while some aqueous samples, such as urine, can be examined directly, although sometimes with considerable uncertainty as to the best indicator to use. Salts of organic acids can be determined by titration with perchloric acid in non-aqueous solution, on the one hand, or by separating the metal on a cation exchange resin and determining the acid produced, on the other, although in this case the metal is only being determined by inference.

Colorimetric methods also have been widely used in determining metals in organic samples without any pretreatment. In some simple, uncomplicated samples, such as acrylonitrile, trace amounts of metal ions can be determined in the material itself merely by adding the reagents, dissolved in a suitable solvent, and measuring the colour produced. For other applications, such as the determination of zinc in urine, better results are obtained by extracting the metal complex into a separate organic solvent phase.

The various other spectrometric methods have also been used for the determination of various elements without much in the way of preliminary treatment. Flame photometry and atomic absorptiometry have probably been the most widely used, and materials such as urine or petroleum products can often be determined by these methods with no more preliminary treatment than a simple dilution. Samples for emission spectrography generally need either to be evaporated on to the electrodes or sampled by one of the procedures for the spectrography of solutions. When suspended matter is present the rotating disc technique has been found very satisfactory.

Direct precipitation of the element to be determined can be employed in some instances where the sample is already a liquid, or where it can conveniently be dissolved in a suitable solvent. This procedure has most commonly been applied to urine, where lead, for example, has often been separated as phosphate or oxalate. A less usual application has been described in which mercury was determined in vinyl esters by dissolving the material in an acid organic solvent and precipitating the mercury as the sulphide. This procedure was found to be more successful than many others which involved much more vigorous methods for the removal of the organic matter.

In addition to the types of sample so far discussed which require virtually no preliminary treatment, there are many more applications which, although they do not involve the complete destruction of the organic matter present, do involve some considerable disruption of the sample before the methods to be used for the final determination can be applied. Such samples are generally more complex than those which are amenable to direct determination, they contain more insoluble matter, and the elements to be determined occur either in small amount or in combined or complexed form, or both. Very many procedures have been described for converting the metals or metalloids in these samples into simple ionic form, but virtually all of them involve treatment with an acid, an alkali, an oxidizing agent or complexing agent or with a combination of these.

Much the commonest are the acid reagents, and of these hydrochloric acid is far and away the most frequently used in this context. It serves the dual functions of acid and complexing agent, the salts it forms with most metals are water soluble, and it has no oxidizing effect upon the reagents that are likely to be used after it. The procedures described vary from shaking in the cold to heating under reflux, and with acid concentrations ranging from a few per cent to concentrated. With organic samples that are not readily dispersed such as ointments or petroleum fractions, it is sometimes desirable to dissolve them first in a solvent such as chloroform, although this is by no means always essential.

With materials containing protein it is often necessary to precipitate the protein as well as to treat the sample with hydrochloric acid, and trichloracetic acid is frequently used for this ancillary purpose. In an investigation into the determination of iron and copper in serum it was shown that treatment with 6 N hydrochloric acid, followed, after 10 min standing, by trichloracetic acid was sufficient to give complete recoveries, but that if hydrochloric acid of higher or lower concentration was used, lower results were obtained. [1]

Other acids are used more rarely, but various applications of acetic acid, nitric acid and perchloric acid have been described. Nitric acid acts on samples both as an acid and as an oxidizing agent, achieving solution by virtue of both properties, while perchloric acid, at least in this context of limited attack, serves primarily as an acid, the low temperatures used restricting its oxidizing power.

Alkaline reagents have been much less commonly used than acids, being mainly restricted to specialized applications such as the hydrolysis of the organic compounds of elements which tend to form insoluble oxides in acid solution, such as niobium and tantalum, or where losses due to volatilization from acid solution are feared. An example which must represent the ultimate in mild treatments of this type is the conversion of the arsenic in primary Lewisite to the inorganic form by refluxing with aqueous sodium bicarbonate solution. [2]

The use of oxidizing agents which are not also acids is also restricted to rather specialized applications. Probably the widest application is the use of bromine for the decomposition of anti-knock compounds in petrols. This has generally been used for the determination of lead, but compounds containing other elements have also been decomposed. This method is less

[1] Venturo, S., and White, J. C., *Analyst*, **79**, 39 (1954).
[2] Fournier, R. M., *Mém. Poudres*, **40**, 385 (1958).

satisfactory with highly unsaturated petrols because of the considerable amounts of bromine consumed. Oxidation with bromine has also been used for the decomposition of organo-mercury compounds, and treatment of blood or urine with chlorine has served to liberate arsenic prior to its precipitation as sulphide.

The use of complexing agents to cause the breakdown of metallo-organic compounds is fairly rare, although their use to bring about the separation of metals already in ionic form is much more common. One valuable procedure which does come into this class is the use of strong sodium sulphide solution for the extraction of organo-mercury compounds, which on acidification break down to give mercuric sulphide. [1]

One final technique which deserves special mention is the use of sulphuric acid and potassium permanganate for the preparation of samples prior to the determination of mercury. This method has been applied most commonly to urine samples, but other materials such as blood, homogenized tissues and apple skins have been handled successfully. The success of the method depends upon the destruction of groups in the organic material which would complex the mercury, and prevent its extraction. The treatment generally employed is boiling under reflux, but the use of a sealed container under slight pressure at 80°C has also been described.

[1] Sporek, K. F., *Analyst*, **81**, 478 (1956).

CHAPTER 3

SOURCES OF ERROR

A GENERALIZED procedure for the determination of an element in an organic matrix would probably involve the following steps:

1. Sampling.
2. Disruption of the sample.
3. Manipulation.
4. Measurement.

Each of these steps can give rise to errors which will affect the accuracy of the final result, although the importance of each will vary with the sample, the element and the method of determination.

1. Sampling

Representative sampling is fundamental to all successful analytical work, and any analysis is, at best, only as good as the sample on which it is based. Errors introduced at this stage can be due to contamination or to loss of material, and each type of error can be caused by a number of mechanisms.

The usual technique used to reduce a possibly non-homogeneous sample to a uniform and finely divided state is rough cutting followed by maceration in some kind of a mill, but this type of cutting and grinding action can easily introduce contamination which is significant in trace level determinations. Relatively early work[1] showed that most grinding techniques, apart from hand grinding with a mortar and pestle, introduced traces of a number of elements into the sample, sometimes in amounts ten times the naturally occurring level. Other interesting facts which arose from the same work were that the level of contamination in the ground sample varied with the nature of the material being ground, and varied also with

[1] Hood, S. L., Parks, R. Q., and Hurwitz, C., *Ind. Eng. Chem.* (*Anal. Ed.*), **16**, 202 (1944).

the size of the sample; the smaller the sample the greater the contamination. The much more widespread use today of stainless steels, and plastic components, has probably reduced the seriousness of the problem, but it undoubtedly still exists, and it is an obvious precaution to choose a cutting or grinding machine which is not likely to be made from a material with a significant content of the elements being determined.

A serious source of contamination with plant material is the presence of soil or spray residues on the various surfaces, and such samples are commonly washed to prevent this extraneous material from being included in the analytical sample. However, this washing can be criticized on the opposing grounds that either it is ineffective in removing surface contamination, or that it removes more than material superficially present, and leaches trace elements from the sample. A number of studies have been carried out to investigate this question, but the results are inconclusive and to some extent contradictory, although it is generally agreed that the use of unwashed plant material samples for trace element analysis will introduce an unacceptable degree of uncertainty into the results.

Various workers [1,2] have found that brief periods of washing in dilute acid or detergent solutions do not cause significant leaching of micronutrients from plant samples, but that the efficacy of these treatments in removing dust or spray residues is variable. Small amounts of iron, copper and zinc in spray residues were found to be readily removable from some plants, but not from those with rough or hairy surfaces; manganese and molybdenum were always retained by the plant to some extent. In this last instance, however, it was suggested that some of the material was absorbed and utilized by the plant, so that it could no longer be considered to be present as superficial contamination. When longer periods of washing have been employed to reduce the degree of contamination even the use of distilled water has been shown to cause losses by leaching. [3]

Many other possible routes exist by which samples may be contaminated, in addition to the few mentioned here, and for critical work a very careful watch must be maintained, to minimize errors arising from this cause.

[1] Arkley, T. H., Munns, D. H., and Johnson, C. M., *J. Ag. Food Chem.* **8**, 318 (1960).
[2] Nicholas, D. J. D., Lloyd-Jones, C. P., and Fisher, D. J., *Plant and Soil*, **8**, 367 (1957).
[3] Tukey, H. B., Jr., Tukey, H. B., and Wittwer, S. H., *Proc. Am. Soc. Hort. Sci.* **71**, 496 (1958).

2. Disruption of the Sample

This is the subject of the bulk of this monograph and will only be mentioned briefly here. Errors arising in this stage can be due either to contamination from external sources, or to losses of material from the sample, both of which are attributable to a variety of causes. Contamination is discussed in the next section, but the losses arising in this stage will be dealt with in much more detail later.

3. Manipulation

Any method of determination involves some manipulation, even if it is no more than the introduction of the prepared sample into a suitable container, and any manipulation introduces the possibility of error, either by contamination or by loss of material. Contamination can arise from reagents, from apparatus, or even from the laboratory environment, while the commonest form of loss, apart from mechanical loss, is by adsorption on one or more of the many surfaces with which the material comes into contact.

Contamination can arise from many sources, not all of which are within the control of the operator. Reagents, including water are often used in considerable quantities, and contamination from this source must be borne in mind. Many of the common reagents are now produced in specially purified grades for particular types of analysis, while normal reagent grade chemicals are adequate for determinations at the rather higher levels. Most suppliers publish impurity limits for the reagents they supply, and in some instances this is reinforced with actual batch analysis data. Laboratory distilled water is generally of adequate quality although water produced by ion exchange, despite its very low ion content, may contain sufficient organic matter to be troublesome in some applications.

It is sometimes desirable to carry out further purification of reagents in the laboratory before use, and this is most readily achieved either by distillation or by extraction. Distillation can be used for materials such as hydrochloric acid, nitric acid, or ammonia, although it is not always easy to improve upon the levels achieved by the manufacturers. Many materials which can be prepared as approximately neutral aqueous solutions can be significantly improved in purity by extraction with organic solutions of complexing agents of which dithizone or 8-hydroxyquinoline in carbon tetrachloride or chloroform are probably the most useful.

One reagent that is used in large quantities in dry oxidations is air, and this can sometimes be a serious source of contamination, the nature of which may well vary with the locality. Some workers have used filtered air for this type of oxidation to remove this hazard, but for the majority of work, down to the part per million level, this precaution is probably unnecessary, unless the atmosphere is known to be heavily contaminated.

Contamination arising from apparatus. The most commonly used material for the construction of laboratory apparatus is borosilicate glass, while silica, platinum and porcelain are also widely used for specific applications. Contamination can be introduced from any of this apparatus either through external contamination of the apparatus itself or through solution of the material of construction during the various manipulative processes.

External contamination is generally removable by efficient acid cleaning of the apparatus before use, and chromic-sulphuric acid is a very widely used and generally successful cleaning agent. Chromic-nitric acid is sometimes preferred for apparatus used in the determination of elements having insoluble sulphates, but both of these reagents suffer from the disadvantage that chromium is readily absorbed upon glassware and silica ware, and is then exceedingly difficult to rinse off. To avoid this difficulty Thiers[1] has recommended the use of mixed nitric and sulphuric acids which he found very satisfactory, while a solution consisting of 30% nitric acid, 5% hydrofluoric acid, 5% Teepol and 60% water has proved excellent for cleaning glass and silica ware. This mixture is not suitable for volumetric glassware.

With regard to the second source of contamination, solution of the apparatus itself, it is evident that the two factors of greatest importance in determining the degree of contamination are the composition of the material, and its resistance to attack by the reagents employed.

The purest material used for the construction of laboratory apparatus is probably silica, for which a purity of greater than $99 \cdot 8\%$ is claimed for normal fused silica ware, and special grades of synthetic fused silica are available in which the non-silica impurities are reduced to less than 1 ppm. In addition to its high purity, fused silica is also a chemically resistant material, particularly at comparatively low temperatures. Nitric, sulphuric and hydrochloric acids are said to have no action on Vitreosil

[1] Thiers, R. E., in Glick, D., Ed., *Methods of Biochemical Analysis*, Vol. 5. Interscience, New York, 1957.

at temperatures up to 1000°C,[1] although a maximum loss of weight of 1 mg/dcm², in hydrochloric acid, is also quoted. Phosphoric and hydrofluoric acids do attack silica, although the attack at low temperatures is said to be slight.[1] Ammonia solution will not affect silica ware, but solutions of alkali hydroxides or carbonates erode the surface relatively rapidly. In the type of work under discussion silica is used most commonly for the construction of apparatus for high temperature fusions or oxidations, and under these conditions many materials will react with it. At the temperatures normally used for the destruction of organic matter—say up to 600°C—the number is much reduced, although lead salts, some halides, alkali carbonates and hydroxides, and a number of other carbonates and oxides will all cause significant attack.

Singer[2] has discussed the effect of various constituents in weakening the silica structure, and if such attack on the surface of the silica has been sufficient, part of the material may be dissolved during subsequent operations, leading to contamination of the solution. However, unless the presence of silica itself is important the high purity of the material makes the effect of this solution of the construction material of reduced significance.

Similar arguments can be applied to the use of platinum apparatus for oxidation reactions. Laboratory ware is generally made from platinum alloyed with a little iridium to improve its stiffness. For use at low temperatures the presence of iridium is of no importance, but loss of weight can occur after ignition at high temperature. Other elements frequently found in laboratory platinum are copper and iron, although the latter is now much less common than it was formerly. The chemical resistance of platinum is very high and few of the materials occurring in organic matrices are likely to attack it at temperatures below 600°C. Nonetheless, care must be taken in oxidations involving readily reducible metals such as copper, silver or gold where there is a risk of reduction to the metal by the carbonaceous material present, and subsequent alloying with the platinum. However, even if some degree of attack does occur, with consequent solution of some of the platinum, the presence of traces of platinum or iridium will usually not be a serious embarrassment.

The third material commonly used for apparatus employed in dry oxidation reaction is porcelain. This is usually a glazed pottery rather than a

[1] The Thermal Syndicate, *About Vitreosil*. The Thermal Syndicate, Wallsend, England, 1958.
[2] Singer, F., *Trans. Brit. Ceram. Soc.* **50**, 265 (1950).

poreclain, with a comparatively readily fusible glaze material coating a much more refractory body. It is primarily the composition of the glaze which determines the chemical resistance of the apparatus, and the nature of the contamination introduced by attack upon it. Sandell[1] mentions the presence of heavy metals such as lead, and other workers[2,3] have reported the extraction of copper in relatively large amounts from the glaze of porcelain crucibles, so this material must be considered to present a greater contamination risk than either silica or platinum.

For oxidations in solution the apparatus used is nearly always constructed of resistance glass, which is generally a borosilicate glass containing a number of other oxides in small amounts. Comparisons of the compositions of several glasses of this type are to be found in many texts and the minor constituents present include the oxides of sodium, potassium, aluminium, magnesium, calcium, barium, zinc, iron, manganese, arsenic and antimony. The quantity, and even the presence, of each of these oxides will vary from one glass to another, and the composition of the glass should be borne in mind when dealing with traces of a particular element. In a report on the determination of zinc[4] several workers specifically mentioned the extraction of zinc from the glassware ". . . even Pyrex".

The chemical resistance of these glasses is greatest towards acid solutions and much less toward alkaline solutions. The data available in the manufacturers literature is rather sparse, but the information in Table 3.1 has been compiled from publications produced by three British manufacturers.

Losses of trace elements by adsorption. In determinations of small amounts of various elements in organic samples dilute aqueous solutions of these elements may be obtained at a number of stages. The techniques used for the destruction of organic matter nearly all produce an acid solution of the element, either by dilution of the acids used for wet oxidation, or by the use of acids to dissolve the ash produced by dry oxidation. With these solutions there is the possibility that some quantity of the element of interest will be bound to the wall of the containing vessel, and if the total

[1] Sandell, E. B., *Colorimetric Determination of Traces of Metals.* Interscience, New York, 1944.
[2] St. John, J. L., *J.A.O.A.C.* **24,** 848 (1941).
[3] Elvehjem, C. A., and Lindow, C. W., *J. Biol. Chem.* **81,** 435 (1929).
[4] Holland, E. B., and Ritchie, W. S., *J.A.O.A.C.* **24,** 349 (1941).

TABLE 3.1. THE DURABILITY OF BRITISH BOROSILICATE GLASSES

	Loss in weight mg/dcm²		
	Firmasil	Pyrex	Phoenix
Hydrochloric acid (constant boiling)	<0·1 4 hr	1 6 hr	2·3 6 hr
Sulphuric acid (4 hr)	0·3 Conc. 135°C	0·5 78–80% 130°C	
Perchloric acid	<0·1 4 hr 135°C		0·1 60% 12 hr boiling
Sodium carbonate (boiling)	40 N 4 hr	40 N 6 hr	65* 0·5 N 6 hr
Sodium hydroxide (boiling)	200 2 N 4 hr	120 N 3 hr	185* N 4 hr
Water	3 4 hr 150°C	1·9 8 hr 121°C	0·1 5 hr 145°C

* From small-scale graph.

amount present is very small this loss could be quite significant. Similarly, when the method to be used for determining the element is being calibrated, standard solutions of comparable concentrations are used, and a similar risk of loss from these exists also.

As far as the accuracy of a determination is concerned, incorrect measurement of the sample and incorrect calibration of the method are equally serious, and it is necessary to determine the extent of these possible losses.

Early work by Leutwein[1] indicated that serious losses of metallic elements occurred upon storage in bottles made of Jena glass. He found that molybdenum, vanadium, titanium, and nickel solutions at a concentration of 10 ppm fell to 20 to 40% of this original strength in 75 days and that gold, platinum, palladium and ruthenium fell to rather lower levels in 230 days. In quartz flasks only palladium showed significant losses. He used both acid and alkaline solutions, and attributed the losses to a combination of base exchange and adsorption.

[1] Leutwein, F., *Zentr. Mineral Geol. A.* 129 (1940).

Since then a considerable amount of work has been carried out, although not all of it is reported in the literature, and the general picture is probably a little clearer. Losses of various elements at low concentrations are still reported. For example, West et al.[1] have shown that serious losses of silver, at the part per million level, are caused by adsorption on to Pyrex, flint glass, polyethylene and silicone-coated vessels from solutions at pH 7, thus supporting earlier work by Pierce. Similarly, losses of strontium from dilute solutions (approximately 0·09 to 90 ppm) at pH 7·3 ranged up to 66%[2] and losses of zinc, copper, iron, lead and aluminium on Pyrex ranged from 1 to 10% after 2 weeks.[3] On the other hand, much work has been carried out to show that, under the right conditions, dilute solutions can be stable for long periods of time. The right conditions, at least for metallic elements, are usually acidic, with the actual minimum acidity required varying from one element to another. For example, it has been shown that while adsorption of manganous ions is negligible below pH 5[4] storage of solutions of four-valent plutonium may need a pH below 2[5] and 0·1 ppm solutions of mercury seem to require N hydrochloric acid for stability in borosilicate glass containers.[6]

In this context it is relevant to consider the light thrown upon this problem of adsorption by the problems facing the suppliers of radioactive materials. Solutions of these materials have two properties that have a bearing on this discussion: they are often prepared at extremely low concentrations, particularly when carrier free, and it is very easy to locate them by means of suitable radiation detectors. The fact that these materials can be stored for long periods in glass vessels without serious loss is a good indication of the efficacy of the conditions used, and this view is further supported by the existence and success of the trade in standardized radioactive solutions. These solutions are despatched all over the world in glass ampoules without any evidence of loss of material by adsorption. When it is remembered that modern methods of absolute counting are accurate to better than 0·1% the importance of this fact is re-emphasized. The conditions used for these standards vary from supplier to supplier,

[1] West, F. K., West, P. W., and Iddings F. A., *Anal. Chem.* **38,** 1566 (1966).
[2] Boocock, G., Grimes, J. H., and Wilford, S. P., *Nature,* **194,** 672 (1962).
[3] Healy, G. M., Morgan, J. F., and Parker, R. C., *J. Biol. Chem.* **198,** 305 (1952).
[4] Benes, P., and Gorba, A., *Radiochim. Acta,* **5,** 99 (1966).
[5] Samartseva, A. G., *Soviet Radiochemistry,* **4,** 572 (1962).
[6] Sion, H., Hoste, J., and Gilles, J., in Cheronis, N. D., Ed., *Proc. Symp. Microchem. Tech. Pennsylvania.* Interscience, New York, p. 959 (1961).

but one well-known organization which lists standards of more than sixty different nuclides works, where possible, to a standard concentration of 25 ppm of the element and 0·01 N nitric acid. Even lower concentrations of elements can be used quite safely, and the laboratories of the International Atomic Energy Agency have shown, for example, that caesium and cobalt are quite stable at 2 ppm provided that the acid concentration is raised to 0·1 N. A point worth noting is that the acid used can have a two-fold effect, acting both as acid and as complexing agent. For example, again from radiochemical practice, it was found that whereas silver 110m, of relatively low specific activity, was quite stable in nitric acid solution, silver 105, when prepared carrier free, was not. The problem was overcome by using hydrochloric acid solution instead, where the silver chloride complex was stable, and the silver concentration was so low that the solubility product of silver chloride was not exceeded.

Generally speaking, it can be said that the majority of elements can be prepared in solution at low concentration provided that the right conditions are employed. For information on the correct conditions the catalogues of some of the major radio-isotope suppliers* contain, as incidental information, the fruits of much experience and investigation, and are often most valuable.

There has been considerable discussion of the relative merits of glass and polyethylene for the storage of dilute solutions, and Thiers[1] has concluded that unplasticized, high-pressure polyethylene is the best material of all. It has the advantage of freedom from contamination—although the same cannot be said of the now common low-pressure material—and it does not appear to be unduly subject to adsorption problems. However, it has been shown that loss of water vapour can occur from polyethylene ampoules,[2] although these losses are unlikely to be serious when considering the accuracy of most trace element determinations. An interesting solution to the combined problems of glass and polythene has been the use of polythene bags as liners for glass bottles.[3] In general, it is probably fair to say that either glass or polyethylene can be used for these dilute solutions provided that care is taken to achieve and maintain suitable conditions.

* For example: The Radiochemical Centre, Amersham, Bucks, England.
[1] Thiers, R. E., in Yoe, J. H., and Koch, H. J., Eds., *Trace Analysis*. Wiley, New York, 1957.
[2] Keith, R. L. G., *Nature*, **196**, 500 (1962).
[3] Healy, G. M., Morgan, J. F., and Parker, R. C., *J. Biol. Chem.* **198**, 305 (1952).

Another important source of possible error is the loss of material on precipitates or other solid material produced or existing in the body of the solution. At the very lowest concentration levels it is possible that adsorption of ions on to particles of dust or of solid substances in colloidal solution might cause losses, but at the part per million level this is not likely to be serious. More serious is the loss by adsorption on precipitates formed in the solution, either intentionally or accidently. The readiness with which precipitates formed in solutions of high pH, will carry down ions present in low concentration is well known, and this property is used in numerous purification and concentration processes. In general it is wise to avoid the formation of such precipitates in solutions containing only traces of ions, and indeed the formation of any precipitate must be viewed with suspicion. Lead can be lost from highly acid solutions if the presence of large amounts of calcium in the sample leads to the precipitation of calcium sulphate, and the formation of any insoluble material should be avoided when possible.

4. Measurement

The errors involved in the final measurement of the element being determined will obviously depend on the method of determination used, and the nature and amount of the element present, and consideration of them falls outside the scope of this work.

CHAPTER 4

WET OXIDATION

THE vast majority of wet oxidation procedures involve the use of some combination of four reagents (sulphuric acid, nitric acid, perchloric acid or hydrogen peroxide) and before considering wet oxidation methods as such, it is useful to consider these four reagents separately.

A. Sulphuric Acid

This is the most frequently used component of wet digestion mixtures, and is widely employed in conjunction with nitric acid, perchloric acid and hydrogen peroxide.

The interaction of sulphuric acid with organic compounds is very complex, and, under suitable conditions, can cause oxidation, sulphonation, esterification, hydrolysis of esters, dehydration, or polymerization of which the most important reactions, from the viewpoint of wet digestion, are probably oxidation and dehydration.

The oxidizing action of sulphuric acid is attributable to the simplified reaction:

$$H_2SO_4 \rightarrow H_2O + SO_2 + 0$$

although in most digestion procedures the function of the sulphuric acid is more to facilitate the action of the other oxidizing agents, than to serve as a primary oxidant itself. The oxidation reaction above requires the consumption of some 10 to 20 g of sulphuric acid for each gram of organic material, the largest amounts being required for hydrocarbons, and the smallest for highly oxygenated compounds,[1] and this type of reaction, leading to the production of large amounts of sulphur dioxide, certainly occurs during wet oxidation with mixtures containing sulphuric acid.

The use of sulphuric acid is the basis of Kjeldahl's method for the

[1] Bradstreet, R. B., *The Kjeldahl Method for Organic Nitrogen*. Academic Press, New York, 1965.

19

destruction of organic matter prior to the determination of nitrogen, although sulphuric acid alone is rarely used. In most procedures, potassium sulphate is added to the acid to raise the temperature, and various catalysts are added to hasten the reaction. Mixtures of this type have found little application for the determination of elements other than nitrogen, probably because the presence of large amounts of salts and catalyst residues is likely to cause interference in the subsequent determinations.

Generally the charred residue produced by sulphuric acid treatment is further treated with another oxidant, such as nitric acid or hydrogen peroxide, which continues the decomposition by further oxidizing the partially degraded fragments.

In addition to its function in partially degrading organic materials by its own action, the presence of sulphuric acid in mixtures containing other oxidizing agents also serves to raise the boiling point of the mixture and so enhance the action of the other oxidants. This has been shown for mixtures of nitric and perchloric acids[1] where the addition of sulphuric acid raised the boiling points of some mixtures by 70 or 80°C (Fig. 4.1).

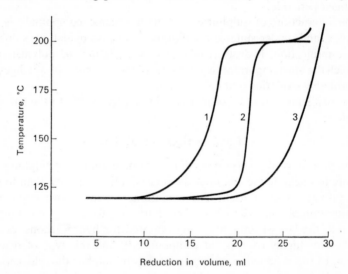

FIG. 4.1.   Effect of sulphuric acid on the boiling point of mixtures of nitric acid and perchloric acids. Curve 1, nitric, perchloric and sulphuric acids. Curve 2, nitric and perchloric acids. Curve 3, nitric and sulphuric acids.

[1] Gorsuch, T. T., *Analyst*, **84**, 135 (1959).

The main disadvantages associated with the use of sulphuric acid are its tendency to form insoluble compounds, and, ironically, its high boiling point.

The insolubility of the alkaline earth sulphates can cause difficulties when dealing with samples rich in these elements, such as bone or milk, difficulties which are particularly serious when recovering lead which itself forms an insoluble sulphate. [1]

The high boiling point of sulphuric acid makes it difficult to remove the excess acid after completion of the oxidation, and while this is not important for elements readily extractable from acid solution, it is inconvenient for those requiring alkaline conditions, as it involves neutralization, with the attendant problems of contamination. The difficulty might be minimized by limiting the sulphuric acid taken to the amount which will be consumed in the digestion, but it is not always easy to judge this.

B. Nitric Acid

Nitric acid is much the most widely used primary oxidant for the destruction of organic matter. It reacts readily with both aromatic and aliphatic organic materials, giving rise to oxidation, esterification and nitration reactions. Aliphatic polyhydroxyl compounds are particularly susceptible to oxidation by nitric acid, being rapidly degraded to simple carboxylic acids, a factor of considerable importance in the destruction of natural materials which contain many such compounds.

The normal concentrated acid boils at about 120°C, a factor which assists in its removal after oxidation, but which correspondingly limits its effectiveness. Nitric acid is commonly used in the presence of sulphuric acid, which partially degrades the more resistant material, or with perchloric acid, which continues the oxidation after the nitric acid has been removed.

The nitrating power of nitric acid can sometimes be a disadvantage, as many nitro compounds are more resistant to attack than their parents. When this constitutes a serious risk it may be desirable either to start the oxidation with dilute nitric acid, which can oxidize some aromatic compounds, or to degrade the material with sulphuric acid before adding the nitric acid.

[1] Gorsuch, T. T., *Analyst*, **84**, 135 (1959).

C. Perchloric Acid

Although this acid has been used for hundreds of thousands of oxidations without incident, the occurrence of occasional explosions of quite stunning violence, has led to it being viewed with grave suspicion.

There is no doubt that, handled with knowledge and care, this reagent is extremely efficient in the destruction of organic material, and from a technical rather than a safety viewpoint, a mixture of nitric and perchloric acids probably has less disadvantages than any other oxidation mixture. However, its degree of tolerance towards mishandling is probably less than is the case with the other common wet oxidation reagents, and for use in unskilled or inexperienced hands avoidance is perhaps better than regret.

This does not in any way imply that an experienced worker with the reagent, who understands its foibles, has any more to fear from perchloric acid than from many of the other reagents in common use. It merely emphasizes that for the newcomer to the field, most of the other oxidation techniques will allow him to make more mistakes in safety.

Reagent grade perchloric acid is generally supplied as either a 60% or a 72% aqueous solution. If the 60% acid is heated it loses water until it reaches a concentration of approximately 72%, when the azeotropic mixture evaporates without further concentration. These fairly concentrated aqueous solutions are stable, and when cold have no oxidizing power. When heated the oxidation potential increases as the temperature rises to 203°C, the boiling point of the acid water azeotrope.

A great deal of work on the applications of perchloric acid has been carried out by G. F. Smith who has carefully defined the conditions under which it can be used without hazard.

Smith has stated that the danger with perchloric acid arises when it is heated with hydroxyl compounds, which give rise to the formation of unstable perchlorate esters. However, hydroxyl compounds such as alcohols and carbohydrates are very susceptible to oxidation with nitric acid, and preliminary treatment with this reagent can be used to remove this source of danger. Accordingly, the use of nitric acid in addition to perchloric acid will give protection against explosion during the oxidation of these hydroxylic materials, provided that sufficient nitric acid is present, and the reaction is allowed to proceed for an adequate period.

There is another class of materials for which preliminary treatment with nitric acid is not necessarily adequate. These are the materials which remain immiscible with the perchloric acid solution, forming a separate phase within the oxidation system so that the reaction with the oxidants can take place only at the interface. This situation appears to be particularly hazardous when the immiscible phase is a fat or wax, rather than a mineral oil, perhaps due to the risk that high temperature hydrolysis of a fat can lead to the production of an alcohol at a temperature where reaction and decomposition would be almost instantaneous. Without being able to cite chapter and verse in support, it is the author's clear impression that of all the fatty materials he has handled, milk fat has probably proved the most dangerous, a situation which, if true, might possibly be related to the short chain length of many of the fatty acids of which it is composed.

Having pointed out the causes from which danger may arise, it is clear that the vast majority of organic materials can be destroyed safely with mixtures containing perchloric and nitric acids, with, in some instances, the inclusion of sulphuric acid also. It is equally clear that to ensure lack of hazard a number of rules must be obeyed. A code of practice for handling perchloric acid has been published by the Society for Analytical Chemistry, [1] and should be studied by intending users. The most important points to watch are:

1. Do not bring materials containing hydroxyl groups into contact with perchloric acid. This includes items such as paper tissues which should not be used for mopping up spills, etc.
2. When oxidizing hydroxylic materials always ensure that adequate preliminary oxidation with nitric acid occurs.
3. Exercise great care when oxidizing immiscible materials. Test a small quantity first, but remember that scaling up can introduce extra hazards.
4. If the oxidation mixture chars, or if there is any sign of difficulty, dilute and cool the mixture at once.
5. There is some evidence that continuous use of perchloric acid in wooden framed fume cupboards can lead to impregnation of the wood, and an increased fire hazard. Cupboards constructed of noninflammable materials are best, but the risk is probably not very great.

The pros and cons of the use of perchloric acid are evenly balanced. Some workers and some organizations will not use it under any circumstances: others claim that where complete freedom from residual organic

[1] Analytical Methods Committee, *Analyst*, **84**, 214 (1959).

matter is essential—as when preparing samples for polarographic investi-gation—there is no other reagent to touch it. My personal attitude in this perplexing matter is that having carried out hundreds or even thousands of oxidations without incident, but having seen the aftermath of one explosion, I would probably avoid its use, other than in exceptional circumstances.

D. Hydrogen Peroxide

Small amounts of hydrogen peroxide have been used for many years as a final treatment to remove small traces of colour remaining in solution after organic material has been oxidized with other mixtures, such as sulphuric and nitric acids. However, within the last decade or so solutions containing 50% or more of hydrogen peroxide, and with a high level of purity leading to high stability, have become available, and these solutions are being used as primary oxidants in association with sulphuric acid. The advantages of such solutions of hydrogen peroxide are well worth considering. They are powerful oxidizing agents: the only decomposition product left behind is water, and their purity is high, with about 1 ppm of sodium, and $0 \cdot 05$ ppm of aluminium being the only metallic impurities: they appear to be safe in use.

Oxidations with hydrogen peroxide in acid solution are considered to involve permono sulphuric acid—prepared *in situ* by reaction with sulphuric acid—a reagent known to introduce oxygenated groups into many kinds of organic molecule. Combined with the dehydrating action of sulphuric acid, this type of reaction will rapidly degrade many organic materials to small species which readily volatilize from the system.

E. Mechanism

A most interesting review has been published by Russian workers concerning the splitting of carbon–carbon bonds.[1] They contend that most wet oxidation procedures are really oxidative–hydrolytic processes, with the insertion of oxygenated substituents by the oxidizing agent serving to facilitate hydrolysis of the organic material. They studied the hydrolysis

[1] Shemyakin, M. M., and Shchukina, L. A., *Quart. Rev.* **10**, 261 (1956).

of many substituted compounds and concluded that in general hydrolysis can occur when the grouping

$$
\begin{array}{c}
\diagdown \quad \diagup \\
C - C \\
\diagup \diagdown \quad | \diagdown \\
\quad OH
\end{array}
$$

occurs, or can arise. This grouping can be formed by hydration of polarized double bonds, and the degree of polarization, and hence the ease of hydrolysis, is affected by the nature of the substitutents on the $\alpha + \beta$ carbon atoms. For example:

$$
\begin{array}{cccc}
\alpha & \beta & \alpha & \beta \\
\diagdown \quad \diagup & & \diagdown \quad \diagup \\
C = C & \longrightarrow & C - C \\
\diagup \quad \diagdown & & \diagup | \quad | \diagdown \\
& & OH \quad H
\end{array}
$$

$$
\begin{array}{cccc}
\alpha & \beta & \alpha & \beta \\
\diagdown \quad \diagup & & \diagdown \quad \diagup \\
C - C & & C(OH) - C \\
| \| \quad \diagdown & & | \quad \diagdown \\
O & & OH
\end{array}
$$

The presence of electron donating substituents on the α carbon, and electron accepting substituents on the β carbon greatly increases the ease of hydrolysis, while the presence of similar substituents on both carbon atoms clearly neutralizes the effect.

When considering this type of mechanism in the context of the normal wet oxidation procedures with mixtures such as sulphuric and nitric acids, and sulphuric acid and hydrogen peroxide, it is interesting to note that sulphuric acid is a dehydrating agent which can readily produce double bonds; that nitro groups are strong electron acceptors and that hydrogen peroxide readily introduces hydroxyl groups into many molecules. When the additional catalytic effect of low pH on the rate of hydrolysis of many partly oxidized compounds is also considered it is clear that the use of the standard mixtures accords well with the princples outlined in the review.

F. Problems

As with the dry ashing methods, problems can arise in wet digestion procedures from two main sources, adsorption on solid material, and volatilization, but the relative significance of the two mechanisms is very different. In general the temperatures involved in wet oxidation methods are very much lower than in dry ashing, and retention losses caused by reaction between the desired element and the apparatus are very much less likely. The same observation applies also to reaction with other solid constituents of the sample, and it is generally considered that this type of loss is of little significance.

One mechanism that is of importance is coprecipitation of the element being determined with a precipitate formed in the digestion mixture, although even this only occurs in some specialized circumstances. The best known example is probably coprecipitation of lead on calcium sulphate precipitates formed when materials high in calcium are digested with mixtures containing sulphuric acid. The formation of such a precipitate should always be viewed with suspicion, and steps taken either to avoid the formation, or to determine whether any coprecipitation does take place.

Turning to the question of volatilization losses, the lower temperatures attained would again be expected to reduce losses of this type, and, the use of a large excess of acid might be expected to convert all forms of the element to a single involatile form, and so prevent volatilization. All of this is true for many elements but there remain a few where volatilization problems still occur.

The simplest, and most obvious case is seen with inherently volatile elements, particularly mercury, where losses occur even at the temperatures reached in wet oxidation. This element requires the use of special techniques which are discussed in the section on mercury.

A slightly more complex situation is seen with ruthenium and osmium, where the volatile form is the most highly oxidized form, the tetroxide. Under the oxidizing conditions found in wet digestions, large losses of the element can occur.

By contrast, some elements can be reduced to volatile forms if reducing conditions are allowed to occur during digestion. As many methods with sulphuric and nitric acids, or sulphuric acid and hydrogen peroxide, involve charring the sample, reducing conditions are readily induced, and losses of selenium have been demonstrated under these circumstances with both of these oxidizing mixtures. For this element there would seem to be a good case for using perchloric acid mixtures, where oxidizing conditions can be maintained throughout.

The last, and probably the greatest, source of trouble is the presence of chlorine in the sample, and difficulties can arise whether it is present in inorganic or covalent form. In general ionic chlorine will give the least difficulty if its presence is recognized and suitable precautions taken. Digestion of the sample with nitric acid will generally remove chloride ion, as nitrosyl chloride, at temperatures below the volatilization temperatures of most other chlorides. When nitric acid is not used, as in sulphuric–peroxide oxidations, acidification with dilute sulphuric acid, and evaporation, should distill off the hydrochloric acid before the temperature rises on charring. However, when the sample contains covalent chlorine, the chloride ion is not produced until the sample is decomposed, a process which generally only takes place at temperatures high enough to volatilize the chlorides as they are formed. Losses of this kind are known to occur with germanium and arsenic, among others, and, unfortunately, similar problems occur in the dry ashing procedures, discussed in the next chapter.

CHAPTER 5

DRY OXIDATION

A. General Discussion

The term "dry oxidation" is generally applied to those procedures in which organic matter is oxidized by reaction with gaseous oxygen, generally with the application of energy in some form. Included in this general term are the methods in which the sample is heated to a relatively high temperature in a stream of air or oxygen; the related low-temperature technique where excited oxygen is used; bomb methods using oxygen under pressure; and the oxygen flask technique in which the sample is ignited in a closed system at approximately atmospheric pressure. Oxidative fusion methods are considered separately.

All the methods involve the following series of processes, although the relative significance of each of them varies from one method to the other.

1. Evaporation of moisture.
2. Evaporation of volatile materials including those produced by thermal cracking or partial oxidation. These may or may not be further oxidized in the gas phase.
3. Progressive oxidation of the non-volatile residue, until all organic matter is destroyed.

Although all three processes occur in all dry oxidations it is not always possible to distinguish them as separate stages. They are perhaps most easily separated in the conventional ashing procedure in which the organic material is heated in an open vessel with free access of air or oxygen.

In conventional analytical practice the first two steps in such a procedure are usually carried out at a temperature much lower than that used to complete the oxidation. This is largely to prevent ignition of the volatile and inflammable material produced by the process of destructive distillation and partial oxidation, as this would lead to an uncontrolled rise in temperature, and an increased chance of loss of material. An exception to this is found with petroleum products where the material is often purposely ignited in order to remove the bulk of the inflammable material.

The preliminary low temperature treatment in the conventional process can be achieved in many ways; by heating gently over a flame, by heating with an overhead source of heat such as an infra red lamp, or even by insertion into the muffle furnace at a low setting. Heating over a flame is rapid, but requires continual attention; the restricted access of air to the interior of a furnace tends to lead to accumulation of inflammable material, with an increased risk of uncontrolled burning: the best method of all is probably the use of an infra red lamp, which, although rather slow, can be left unattended for long periods of time, and, if necessary, lends itself to enclosure of the sample in an atmosphere of filtered air. An apparatus of the type shown in Fig. 5.1 has been recommended by Thiers[1] for this purpose, and has the additional refinement of a hot plate to increase the rate of oxidation and so reduce the duration of the treatment required.

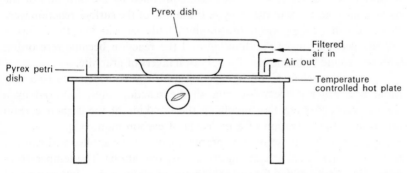

Fig. 5.1   Apparatus for preliminary charring of samples.

A possible alternative method of treatment which does not seem to have been used in this context, would be to carry out the initial treatment in an inert gas atmosphere. This would clearly require the use of an enclosed system, but it might offer some advantages, such as freedom from fear of ignition and the ability to use fairly high temperatures to achieve rapid thermal degradation of the organic material. When the evolution of

[1] Thiers, R. E., in Yoe, J. H., and Koch, H. J., Eds., *Trace Analysis*, p. 637. Wiley, New York, 1957.

volatile material was complete, air or oxygen could then be introduced to carry out the final oxidation of the charred residue.

The third stage of the operation, the progressive oxidation of the non-volatile residue, is a complex process which has not proved amenable to detailed study. The material remaining after the preliminary treatment is a more or less porous mass of charred organic matter with greater or lesser amounts of inorganic material distributed throughout it. The kinetics of the oxidation of such a material will be dependant upon the nature of the material itself, the inorganic substances it contains, and also upon its particle size and porosity. Some information is available from studies on pure graphite, in which the effect of variations in temperature and oxygen pressure on the rate of oxidation has been considered, and in one investigation it was found that, at temperatures up to 800°C, the rate was proportional to the square root of the oxygen pressure. It was shown that this finding was in accord with the model of a very rapid zero order surface reaction between oxygen and carbon, with an activation energy of about 80 kcal/mole, which was rate controlled by the diffusion of the oxygen into the pores of the sample. If the rate of the surface reaction were slowed down, or the pores in the graphite made very shallow, the diffusion process was no longer rate-limiting and the reaction became zero order. This was found to be the case for very thin layers of graphite.

The actual mechanism of the reaction itself was considered to be a very rapid first order process covering all the reaction sites, followed by a slower zero order reaction involving the breaking of the carbon–carbon bonds, and the liberation of a molecule of carbon monoxide.

Findings such as these, on pure graphite, can only be applied with caution to the complex chars existing in dry ashing operations. The temperatures generally recommended for dry ashing are, at about 500°C, low compared with those reported for graphite oxidation, but the chars produced are probably far more reactive, and the possible catalytic effects of the inorganic constituents of the material are unknown.

The large amount of work published on the oxidation of materials such as coal, coke and charcoal is probably more relevant to dry ashing practice, but the increased complexity makes the results more difficult to interpret. There is clearly scope for almost unlimited work on the kinetics of dry oxidation reactions, but this is probably not justified, at least from the practical viewpoint, for the probably limited advances in technique that it might bring.

B. Methods of Dry Ashing

During the oxidation process the element to be determined will behave in one or more of a number of ways.

Ideally it will remain quantitatively in the residue remaining after oxidation, and in a form in which it can be readily recovered, generally by solution of the ash in dilute acid.

Some part of the element may occur in, or be converted to, a volatile form which may escape from the vessel used to contain the organic sample.

Some part of the element may combine with the vessel used to contain the sample, with some component of the inorganic residue remaining after oxidation, or with some other solid material present, in such a way as to be irrecoverable by the normal procedures used for solution of the ash.

The probability that the types of behaviour listed above will occur will vary markedly with the element to be determined and the method of oxidation selected, but whereas, with one technique, care might be taken to prevent a particular type of behaviour, with another the same type of behaviour could be an advantage. This is particularly seen in the case of volatile materials where with open systems it is necessary to prevent loss of material by volatilization, while with the oxygen flask method it is very convenient for the significant element to be volatilized completely.

Generally speaking the division between closed systems and open systems is a real one with the former being best equipped to handle small quantities of material, and the latter being necessary when large samples must be destroyed, particularly for trace analysis.

1. Closed Systems

The classic combustion train, used for the elementary analysis of organic compounds, can best be regarded as a closed system, even though a gas flow is maintained through the apparatus during the combustion. It has been applied to the determination of hetero-elements in organic compounds, and has been used successfully for both involatile materials— which remain as a residue in the combustion boat—and for volatile elements such as mercury which can be distilled out and trapped. However, it is only for volatile materials that it is worthwhile using a system that approximates to a closed system, with effective traps, and the application of such a combustion train is probably best limited to them.

The oxygen flask combustion procedure is both simpler and more flexible than the conventional combustion train, and has been widely used as a preliminary to the determination of both volatile and non-volatile elements. It was originally introduced as a macro-method, but in recent years it has been used almost entirely for small scale work. The principle of the method is very simple. A small sample of the material to be analysed is wrapped in filter paper to give a small packet with a protruding wick (see Fig. 5.2, (a) and (b)). The packet is held in a platinum sample holder mounted on a platinum wire set into a ground glass stopper. A small volume of suitable absorbing solution is placed in the bottom of the flask

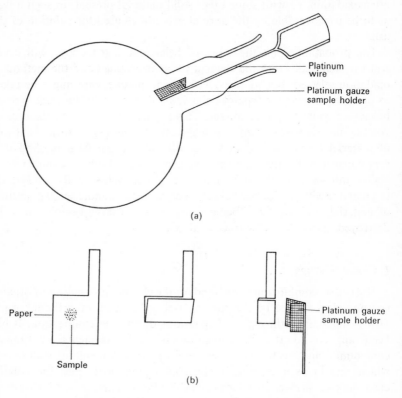

FIG. 5.2   Apparatus for oxygen flask combustions.

which is then flushed with oxygen. The tip of the wick is ignited in a flame, and the stopper inserted rapidly into the flask held horizontally and the flask tipped slightly downwards so that the absorber solution provides a seal. When the smouldering wick enters the flask it ignites in the oxygen atmosphere and the whole sample is consumed in a few seconds: gentle movement of the flask ensures that the last traces of organic material are consumed. The flask is left to stand for 10 minutes, and then shaken vigorously for one or two minutes. The stopper and sample holder are removed and washed down, and the solution containing the element to be determined is ready for measurement.

The oxygen flask method is suitable for the determination of many elements, both volatile and non-volatile, but a few have given trouble. Arsenic, for which the technique might seem ideal, is not really suitable. It attacks the usual platinum holder, and although silica sample holders have been used instead, these are not really satisfactory. Lead and bismuth form alloys with platinum sample holders, and the readily reduced elements also demand caution. Some metals, such as nickel and gallium form insoluble oxides which necessitate special treatment before determination, but many elements such as the alkaline earths, and zinc, cadmium and mercury can readily be determined. The method is indeed very suitable for mercury, an element which causes more trouble than any other with other methods of sample decomposition. It is interesting to note that for mercury the procedure offers such advantages that it has been scaled up again to deal with gram-size samples.

The other entirely closed system that might be mentioned is the oxygen bomb, in which the sample is ignited in a small pressure vessel with oxygen under considerable pressure. It has been used for the decomposition of organo-metallic compounds but not extensively.

2. Open Systems

Much more commonly used are the open systems which depend on the assumption that no significant amount of material will be lost by volatilization. Such methods are commonly employed for the decomposition of organo-metallic compounds, and are almost exclusively used when large samples are handled. Although serious volatilization loss must not occur if open ignition is to be used successfully the possibility of such occurrence must always be borne in mind when considering a projected application.

The conventional dry ashing procedure proceeds via the three stages described at the beginning of this chapter, and provided that there is no gross failure of technique, such as allowing the sample to ignite, it is probable that the majority of the losses occur during the final stage. This may not be entirely true when losses are attributed to reduction processes, as there is some evidence that the charring stage, with its copious evolution of gaseous decomposition products, can be effective in causing reduction: however, in most instances, the losses occur during the high temperature third stage.

The actual temperatures used for the final ignition of the sample vary very widely. For the excited oxygen method described later the sample temperature is said not greatly to exceed 100°C, while for conventional dry ashing, temperatures as high as 1000°C have been used, though very rarely; 450 to 550°C is by far the most commonly used range.

(a) *Volatilization losses.* In considering the likelihood of losses occurring by volatilization it is necessary to consider a number of factors. The first, and most obvious, is to consider whether the element to be determined is present in an inherently volatile form. Some elements, such as mercury, are volatile at the usual ashing temperatures, almost regardless of the chemical form in which they occur, while others are volatile in some forms but not in others. Lead for example, can safely be ashed if it occurs as the sulphate, the nitrate, or the oxide, but if present as the chloride considerable care is required. Plumbous chloride melts at 501°C and at virtually any temperature above this, its vapour pressure is sufficient to introduce the possibility of significant losses when it is present in very low concentration.

The second factor to consider is the possibility of the element reacting with some inorganic constituent of the sample to produce a volatile species. An example of this is the volatilization of elements such as antimony, zinc and lead when the nitrates are heated in the presence of ammonium chloride, due to the production of the corresponding volatile chlorides. Similarly, zinc has also been shown to volatilize if its nitrate is heated with calcium or magnesium chlorides, due to the same cause. However, in no instance has a loss of this kind been demonstrated by reaction with sodium chloride. Indeed, although such a mechanism has been suggested to explain the loss of elements such as iron, the thermodynamics of the necessary reactions are so unfavourable that the reaction of sodium chloride with the compounds of virtually any of the elements of interest in this context, to give volatile chlorides, is improbable in the extreme. However, the presence of sodium chloride can give rise to very serious retention losses in some instances, and the probable mechanism for this is discussed in more detail later in the chapter.

The third factor of importance in assessing the probability of losses by volatilization is the extent to which the element of interest will react with the organic matter of the sample to produce a volatile form. This type of mechanism has been suggested to explain a number of observed losses, such as the very large loss of lead during the ashing of polyvinyl chloride which again was attributed to the formation of lead chloride, and the loss of cadmium during dry ashing which has tentatively been attributed to reduction to cadmium metal by the partly ashed organic matter. Although this last mechanism has not been clearly demonstrated, the free energy change for the reaction of cadmium oxide with carbon is probably just favourable: by contrast the reaction of zinc oxide with carbon to give zinc metal, which has been postulated to explain losses of zinc during dry ashing is certainly not feasible at normal dry ashing temperatures.

(b) *Retention losses.* The other likely cause of losses during dry ashing procedures is the reaction of the element to be determined with some of the solid matter present in the system. In order for a reaction of this kind to constitute a problem it is necessary first of all for the reaction to occur to a significant extent, and secondly for the product of the reaction to be stable to the reagents generally used for dissolving the ash produced, so that the element is effectively prevented from interacting with the detecting agency. The solid matter available for reaction is generally the material of the ashing vessel and the constituents of the ash of the sample itself, and it is obvious that their nature is going to have a considerable effect upon the losses.

The ashing vessels most commonly used are made of silica or porcelain, with platinum as a probable third material. Vitreous silica is a glass, consisting almost entirely of SiO_2, whereas the glaze on porcelain ware is a more complex material containing other oxides in addition to silica. For both these materials the obvious reaction is between the oxide of the element of interest, and the ashing vessel to produce a complex silicate, so causing a loss. This type of reaction clearly occurs, but the ease with which it occurs varies with many factors. Some oxides react much more readily than others, with lead oxide one of the most reactive. It has been suggested that this is connected with its high polarizability which facilitates the approach of the oxide molecules to the active sites in the silicate lattice. Other oxides such as those of calcium or magnesium are, by contrast, considered very unreactive. Similarly, even if silicates are formed, some will be stable to acid attack, while others will readily be decomposed, and so cannot be considered to cause losses. Copper silicate, for example, is very readily decomposed by acids, and even if it were formed, it should not cause retention of copper on the vessel, yet copper is one of the elements most commonly quoted as being troublesome in this respect. In an attempt to explain this apparent paradox it has been suggested that the loss of copper is due, not to the formation of copper silicate by reaction of the oxide, but to reduction of the oxide to copper metal, and diffusion of this into the silica lattice. Some support for this hypothesis can be found in the report that losses of copper were much higher when copper tracer was heated in silica crucibles in the presence of organic matter than when it was heated alone, even though the organic matter might be expected, at least in part, to separate the two reactants and so reduce the retention. Similar difficulties might be expected with other readily reducible elements such as silver, gold and the platinum metals. This leads directly to consideration of platinum ashing vessels, where it might be surmised that losses by reactions of non-reduced oxides would cause no trouble, but that losses due to reduction of the oxides to the elements would be as serious as ever. There is some evidence to support this view.

Reactions involving the inorganic constituents of the ash generally follow very similar lines, as the most significant constituent of the ash is usually silica, and very serious losses have been reported due to the formation of stable reaction products. However, the presence of sodium chloride in some organic materials, particularly those of biological origin, does present a new problem.

Earlier in this chapter it was stated that the presence of sodium chloride does not lead to increased losses of elements by volatilization, but that it does lead to increased retention losses. This is certainly the case when ashing vessels made of silica, and probably porcelain, are used, due to a marked weakening of the silicate structure by chloride attack, leading to greatly increased reactivity and correspondingly increased losses. Examples have been quoted where the presence of sodium chloride reduced the recovery of lead from 80% to 5% under identical conditions.

The serious consequences of this type of reaction, together with the volatilization losses from materials such as PVC mentioned earlier, necessitate careful consideration of the methods used to ash materials containing significant amounts of chlorine, whether it is present in ionic or in covalent form. If the former then losses might occur by either volatilization or retention, depending on the circumstances, if the latter then probably by volatilization only. When chloride ion is present some improvement can sometimes be obtained by acidifying the sample with dilute sulphuric acid, and driving off the hydrochloric acid so formed. In doing this it is essential not to allow the temperature to rise to a point where the chloride of the element to be determined is volatile, bearing in mind the effect of its valency state. The use of well-diluted sulphuric acid helps to keep the temperature low, and it may be advisable to redilute, and re-evaporate the acid to ensure that all the chloride ion has been removed. With materials containing covalent chlorine, the situation is much more difficult, as the relatively high temperatures required to decompose the organic matter ensure that the hydrochloric acid produced as a decomposition product is produced at a temperature high enough to cause the chlorides produced *in situ*, to volatilize.

Another retention problem found during dry ashing, though of a very different kind, is that posed by the sequestering action of some materials produced during ignition. The best known example is the binding of iron by condensed phosphates produced by the action of heat on simple phosphates present in the sample. The generally accepted solution to the problem is to reverse the process by hydrolysing the complex phosphates back again by heating in acid solution.

(c) *The use of ashing aids.* Very many of the reported ashing methods describe the addition of some extra inorganic material to the sample to improve the procedure. These added materials are generally called ashing aids, and they serve one or both of two purposes: to facilitate the decomposition of the organic material, or to improve the recovery of the element to be determined. Very many different materials have been used as ashing aids, and although they can all be allocated to one or both of the categories noted above, the mechanisms by which they achieve these ends can vary.

The most commonly used auxiliary oxidant, used purely to hasten the oxidation of the organic material, is nitric acid. This is generally added towards the end of the ashing process to remove small amounts of carbonaceous material remaining. It has the disadvantage that if it is added too soon, when appreciable amounts of organic material are still present, it can cause the residue to ignite when returned to the furnace, with possible loss of material, but when used with care it can lead to a saving in time, and the production of a clean ash.

Some substances serve as auxiliary oxidants as well as serving other purposes. These are commonly the nitrates of light metals such as magnesium, calcium and aluminium which decompose on heating to give oxides of nitrogen. Of all the ashing aids quoted, these are probably the most widely used.

As well as acting as auxiliary oxidants the light metal nitrates, together with many other compounds, fulfil an important function as inactive diluents. As the organic material in a sample is progressively decomposed, the elements being determined are brought into closer contact with the material of the ashing vessel and the other constituents of the sample ash. If reaction with them is feasible then the increased proximity will increase the chances of it occurring. Under these circumstances dilution of the ash with an inert material such as magnesium oxide should greatly reduce the possibility of undesirable solid state reactions, and so improve the attainable recoveries. The use of relatively unstable nitrates such as those of aluminium or magnesium therefore offers the advantages both of more rapid oxidation and of decreased retention.

These diluting materials improve recoveries by separating potentially reactive substances in the samples but without entering into any reaction with them itself. Another group of ashing aids achieves the same end by altering the chemical nature of some of the constituents. A simple example is the use of sulphuric acid to convert volatile lead chloride to involatile lead sulphate, or to remove chloride ion by conversion to hydrochloric acid. Similarly sodium hydroxide has been used to precipitate iron as the hydroxide or to convert zinc to a zincate. On a rather more hypothetical level the reported effectiveness of boric acid as an ashing aid for the recovery of lead, can also be fitted into the same pattern. In a small number of experiments it has been shown that boric acid not only allowed good recoveries of lead nitrate tracer after heating at 650°C, but also gave quantitative recovery of lead chloride under the same conditions. In view of the volatility of lead chloride this was a somewhat surprising result which was tentatively explained by the formation of a boron–silica–lead glass, which was stable enough to prevent volatilization of the lead, but, because boric oxide tends to form only a two-dimensional network, broke down under acid attack to yield the lead quantitatively into solution. By contrast, the addition of phosphoric acid, which was also effective in preventing the volatilization of lead chloride, caused a very large retention loss. This was explained on the grounds that while P_2O_5, like B_2O_3 is a network former, it is capable of forming three-dimensional networks much more resistant to acid attack, so that the glass formed would not break down under the conditions used to dissolve the ash. The use of boric acid as an ashing aid does not seem to have been investigated in any detail, and its effectiveness in preventing loss of elements other than lead might make an interesting study.

C. Oxidation with Excited Oxygen

A procedure was described[1] in 1962 which differed appreciably from conventional ashing methods in that electronically excited oxygen, produced by subjecting a stream of oxygen gas to a high-frequency electrodeless discharge, was used to decompose the organic matter. The discharge produced reactive atomic and ionic species in the oxygen at low pressures—about one torr—which reacted with the organic material without raising the temperature above about 150°C. The advantages of such a procedure are obvious when the usual causes of low recoveries are considered, and the results quoted show that selenium was recovered to the extent of 99%, and even mercury to the extent of 92%. These are very encouraging figures, and the method would appear to have much to be said for it.

Gleit, C. E., and Holland, W. D., *Anal. Chem.* **34**, 1454 (1962).

CHAPTER 6

OXIDATIVE FUSION AND OTHER METHODS

A. Oxidative Fusion

Methods falling into this category have frequently been described for the decomposition of organo-metallic compounds, and involving, as they do, treatment of the organic sample with an oxidizing melt at fairly high temperatures, they represent some of the most rigorous conditions employed. Most commonly the oxidation is carried out in a completely closed system, with a variety of oxidizing agents being used. The most popular process is probably fusion with sodium peroxide in a Parr bomb, and this has been accepted as satisfactory for a wide range of samples and elements. Other oxidants have also been used, although often in combination with sodium peroxide, and of these potassium nitrate and potassium perchlorate are probably the most common.

For some samples pure, readily oxidized, organic compounds are added to facilitate the reaction, and various carbohydrates, ethylene glycol and benzoic acid have been used in this context.

Despite the severity of the oxidizing conditions occurring in these melts problems can still occur if the organic sample is not intimately mixed with the oxidant. Volatile compounds and waxy solids can cause difficulties in this respect, but rotation of the bomb while the contents are molten will help to reduce the error.

Open vessels have also been used for fusions, although more rarely because of the increased risk of volatilization losses. However a detailed radiochemical study has been published, reporting the recoveries of twenty-six elements after decomposition of organic matter by fusion with a mixture of sodium and potassium nitrates. [1] The procedure described involved adding a twenty-fold excess of an equimolar mixture of the nitrates to the organic sample in an open vessel, together with a little water, and heating

[1] Bowen, H. J. M., *Anal. Chem.* **40,** 969 (1968).

carefully to 390°∓10°C. Charring commenced at 236°C and oxidation was complete in a few minutes at 390°C. The method has been applied successfully to such diverse materials as polyethylene, wheat flour and animal blood, but the major investigation was carried out with kale. A summary of the results for the twenty-two elements relevant to this monograph is given in Table 6.1.

TABLE 6.1. RECOVERY OF ELEMENTS AFTER OXIDATIVE FUSION.

	Recovery	Element
I	Greater than 99%	Silver, Gold, Cadmium, Cerium, Cobalt, Copper, Iron, Manganese, Zinc, Molybdenum, Sodium, Ruthenium, Strontium
II	95–99%	Arsenic, Chromium, Caesium, Iridium, Rhenium, Antimony, Thallium
III	Less than 95%	Selenium (93·7%), Mercury (2·3%)

The recoveries quoted are, on the whole very good, and the method itself is very rapid. However the need to use a twenty-fold excess of the mixed nitrates is a distinct disadvantage in trace element work, both from the small size of sample that can be handled, and from the risk of contamination from impurities in the melt. The large excess of reagent is determined largely by the risk of explosion when very much smaller ratios are used.

The decomposition of the nitrates begins at temperatures of 380°C $(NaNO_3)$ and 400°C (KNO_3) according to the reaction

$$2 MNO_3 \rightarrow 2 MNO_2 + O_2$$

and the oxidation is carried out presumably by both oxygen and oxides of nitrogen.

The residue remaining is then alkaline, and this is believed to be important in the retention of potentially volatile elements such as arsenic, antimony and ruthenium.

This method could easily be of value in determining many elements in organic compounds, particlarly if it could be shown that the recoveries could be maintained in the presence of large proportions of chlorine, both ionic and covalent.

B. Oxidation in an Atmosphere of Nitric Acid

A number of reports have occurred in the literature in which samples have been oxidized in an atmosphere of nitric acid vapour. Generally speaking relatively low temperatures of 300 to 350°C have been used, with a laboratory hot plate as the source of heat.

The apparatus required is very simple, merely a flat bottomed flask containing a mixture of nitric and sulphuric acids, connected to a boiling tube, or other small vessel containing the sample. The connecting tube from the acid flask passes well into the sample container so that the sample is directly in line with the flow of nitric acid vapour. By mounting both the acid flask and the sample container on the hotplate the temperature of the sample is raised, a flow of nitric acid vapour is generated and the sample is very rapidly oxidized. Although the temperature of the hotplate is relatively low the rapid oxidation that occurs leads to a rapid rise in sample temperature and so far little is known of the behaviour of trace elements under these circumstances. A detailed investigation of the recovery of a wide range of elements after oxidation of samples by a procedure of this type would be interesting and valuable.

C. Oxidation with Ozone

The decomposition of organic samples by oxidation with ozone is an area of study which would appear to have been almost totally neglected, even though it might offer an alternative to the excited oxygen method for the low-temperature oxidation of organic samples.

CHAPTER 7

METHODS OF INVESTIGATION

A. General Discussion

In assessing the value of any method of mineralization of organic matter, in any particular set of circumstances, there are two requirements to be considered. Firstly, the method must destroy the organic matter effectively and within an acceptable time, and secondly the element to be determined must remain available, in its original quantity, ready for subsequent determination.

The satisfaction of the first requirement can, on the whole, be recognized fairly readily. Incomplete oxidation can often be seen from simple inspection, and the time required for any degree of destruction is readily measured. For these reasons it has been a relatively simple matter to develop a considerable number of methods which will rapidly and efficiently reduce most types of organic and biological materials to a state in which the organic matter will not interfere with the necessary determinations.

However, when it comes to the question of ensuring the quantitative availability of the element being determined it becomes very much more difficult. For many organo-metallic compounds pure samples of known composition are readily available, and it is merely a matter of weighing the original material, and accurately determining the residue; but when the important element is present only in small quantities, and it bears no definite relationship to the mass of the sample, then the problems multiply. The situation is, in many ways, a classic vicious circle, in which the quantity of the element cannot be determined until the recovery is known, and the recovery cannot be determined until the quantity present is known.

The technique widely used to obtain recovery information in this difficult field is the method of standard additions in which known small additions of the element are made to the sample, the sample is mineralized, the element determined, and the amount of material determined compared with the amount known to have been added. This procedure has provided a large amount of very valuable information, but it does suffer from some serious drawbacks.

1. It is known that the chemical form of the element present in the sample can greatly influence the behaviour of that element during the oxidation procedure, so that, as there is often no guarantee that the tracer added is in the same chemical form, there can be no guarantee that it will behave in a similar manner. This point has been clearly demonstrated by dry ashing cocoa after the addition of lead tracer as either the nitrate or the chloride when it was found that, at an ashing temperature of 650°C, recovery in the former case was 94%, and in the latter only 46%. This is an extreme case, perhaps, but it clearly demonstrates the dangers of assuming that the behaviour of the tracer necessarily parallels the behaviour of the element naturally present in the sample.

2. In most investigations, determination of the element of interest, after the oxidation step, requires some further manipulation of the sample. Errors in the final result can sometimes arise during these steps, but these errors cannot be separated from those introduced by the oxidation technique, so that the uncertainty of the final result is increased.

3. During the course of the whole process, from sampling to determination, errors of two kinds are possible. On the one hand losses can be caused by a variety of mechanisms, while on the other contamination can occur from reagents, apparatus or other sources. These two types of error tend to cancel each other out, so that even an apparent recovery of 100% may be due to the fortuitous compensation of errors, rather than to an error-free technique.

During the last twenty or so years, the wide availability of radioactive materials of high purity and specific activity, has given workers in this field a new technique, which, although similar to the inactive standard addition method offers some notable advantages. The method is the radioactive standard addition method, and it is possible, today, to obtain radioactive nuclides of most of the elements discussed in this book, which have suitable half lives and radiation characteristics to allow them to be used conveniently for this purpose. A list of nuclides and their properties is given in Appendix 1.

The radioactive addition procedure consists of the addition to the sample, before mineralization, of tracer material containing a known amount of radioactivity measured in any convenient units, oxidation of the sample, and determination of the radioactivity remaining using the same units as before.

It must be clearly stated that this method still suffers from the first disadvantage of the inactive standard addition method, that the chemical form of the tracer may well differ from the chemical form of the element in the sample, so that their behaviours during oxidation may well differ also. Under these circumstances it must be accepted that unless particular steps are taken to convert both tracer and traced to the same chemical form, there is a risk that the experiment may be measuring only the recovery of the element when it is present in the chemical form of the tracer, and that extrapolation to other chemical forms may well be unfounded.

However, even with this basic disadvantage, there can be no doubt that the use of a radioactive tracer in recovery experiments of this kind brings about great improvements in their accuracy and reliability.

The measurement of radioactivity can usually be carried out directly on the solution remaining after wet oxidation, or on the acid extract of the ash obtained in dry oxidations. This means that the subsequent manipulation is reduced to a minimum, the risk of introducing other errors, to confuse the picture, is greatly reduced, and the effect of the oxidation stage itself can be considered substantially in isolation. Similarly, the final determination is of radioactivity, and as any contamination of the sample during the various operations will be with the inactive form of the element, it cannot affect the final result. It is, therefore, possible to determine the losses which occur during an oxidation, without the confusion caused by concurrent contamination.

In addition to these two features which show that the radioactive recovery method does not suffer from some of the disadvantages of the normal standard addition method, there are some additional factors which have no counterpart in the inactive method. Much radiation, and particularly electromagnetic radiation such as gamma-rays and X-rays, is of a penetrating nature. This means that radioactive substances which may be fixed tightly to insoluble material such as crucibles or precipitates can still often be identified, and frequently determined. This means that it is possible to distinguish readily between material lost by volatilization and that which is tightly retained by some solid part of the system, and allows more complete material balance sheets to be compiled for an experiment. Another clear advantage of the tracer method is its speed which is often many times faster than the traditional method.

Although the radioactive recovery procedure is the most generally useful, and certainly the most widely used, of the radioactive techniques, it does still suffer from the disadvantage that the tracer and the natural element are not necessarily present in the same chemical form. This drawback can be overcome, to a considerable extent, by the use of neutron irradiation to produce the radioactive form of the element *in situ*, provided that its nuclear characteristics are satisfactory.

Many radioactive nuclides (see Appendix 1) produce, during their decay, gamma-rays and X-rays in constant abundance and of constant energy. By the use of suitable detectors and electronic sorting equipment, gamma-rays of different energies can be separated and measured, and the number being detected used as a measure firstly of the amount of the radioactive nuclide producing them, and, from that, of the element from which the nuclide was produced. As gamma-rays are highly penetrating, this analysis, which is known as gamma spectrometry, can often be carried out on an intact sample.

It can therefore be seen, that if a sample of organic material is irradiated with neutrons some of the element of interest will be converted to its radioactive form, and, under suitable conditions, the amount of element present can be determined by gamma spectrometry on the intact irradiated sample. If the sample is then subjected to an oxidation procedure the fate of the element can be determined by gamma spectrometry on all the fractions produced, and there cannot be the same uncertainty about the behaviour of the sample and the tracer.

This method clearly has some notable advantages over the other methods of investigation, but unfortunately it has some prominent disadvantages also. It will only work for elements of suitable nuclear characteristics, in terms of half-life, cross-section, gamma-ray energy and abundance; for adequate sensitivity, it will generally require access to a nuclear reactor, and, to be able to use it without suffering too much interference from other radioactive nuclides produced during the irradiation, a considerable quantity of expensive electronic equipment is required.

Even these disadvantages can, however, be overcome by the use of another technique, in which the individual tracers are incorporated into plant or animal materials. This can be done by injecting the tracer into animals, or by growing plants, or aquatic animals, in labelled nutrient solutions and allowing time for the radioactive material to be metabolized, so that it occurs in the natural chemical form for that specimen. The amount of tracer present in a sample can be determined by total counting, without the need for gamma spectrometry, as all the radioactivity present must arise from this one source. Measurements of recoveries and losses on the application of different ashing methods can then be interpreted without the normal hesitance due to doubts about the chemical form. This technique is clearly a considerable advance, but it depends, of course, on the availability of suitable facilities for rearing plants and animals, and is less readily applicable to other, non-living natural materials such as oil, textiles or plastics. None the less an increasing volume of information arising from work of this kind is to be hoped for.

Despite more than twenty years of experience with the laboratory use of radioactive tracers, there is still a prevalent feeling that they are dangerous, exotic things, whose only place is in a highly sophisticated, carefully protected research laboratory. This is far from the truth. Care is indeed required in handling radioactive materials, but care is an essential ingredient in all research or analytical work of any value. The amounts of radioactivity required for these tracer studies generally lie in the microcurie* or tens of microcuries region, and these quantities can be handled in any reasonable analytical laboratory. It is necessary to take certain precautions, but these are largely a matter of common sense, and are not particularly onerous; they certainly do not involve lead or concrete shielding, or remote handling. Similarly, the basic counting equipment necessary for such investigations, can be purchased for sums measured in hundreds of pounds rather than thousands, and is certainly cheaper than many items of equipment found today in quite humble laboratories.

Information regarding these matters is available from many sources, including text books and government publications. A useful starting point is the *Radiochemical Manual* (50/-) published by The Radiochemical Centre, Amersham, Bucks, England, who also provide a useful range of free review booklets covering many applications of radionuclides.

Throughout this book much emphasis is placed on the results obtained from radiochemical investigations of the problems associated with the destruction of organic samples, as it is believed that these offer the best methods at present available for unravelling their complexities.

* One microcurie = $3 \cdot 70 \times 10^4$ nuclear transformations per second.

B. Choice of a Radioactive Tracer

The advantages of using radioactive tracers to study the losses occurring during the oxidation of organic materials may appear to be fairly clearcut, but if a worker new to the field wishes to make use of the technique, he may have some problems in selecting the best tracer to use. A glance at the *Radiochemical Manual* reveals tables containing details of hundreds of different nuclides, with, in some cases, perhaps eight or nine isotopes of a single element. Deciding which of these best fits the requirements of the experiment may seem a very difficult task, and it is the purpose of this section to try to simplify it by explaining the principles on which selection should be made.

Before considering this further, however, it is useful to consider the methods available for detecting radiation, as the type of equipment available may determine the tracer to be used, or, conversely, the desirable tracer may require a particular type of counting equipment.

Basically there are two types of detector, ionization detectors and scintillation detectors. For tracer work the Geiger counter is the only ionization counter of importance, and the detector for this is available in two main forms (see Fig. 7.1). A detector for handling liquids is shown as (a); in this the liquid sample fills an annular cup surrounding the sensitive volume of the detector. The radiation from the solution must pass through the liquid, and through the glass envelope surrounding the sensitive volume, and as the glass must be fairly thick for mechanical reasons, fairly energetic radiation is required. The detector is not very sensitive to gamma photons, so it is best used for counting beta particles of fairly high energy, say over 0·75 MeV. However, as the sample is presented to the detector in solution the geometrical relationship between sample and detector is very reproducible, and provided that the density of the solution does not alter much, the measurements are also very reproducible. With the detector for measuring solid samples (Fig. 7.1, (b) the sample and the sensitive volume are separated only by an airspace and a thin mica window, so that beta particles of lower energy can be measured. However, the geometrical reproducibility is somewhat less satisfactory, particularly for low energy beta particles where small variations in the thickness or distribution of the sample can greatly influence the counting efficiency; if steps are taken to overcome this by "infinite thickness" counting, the overall sensitivity can be very low. For

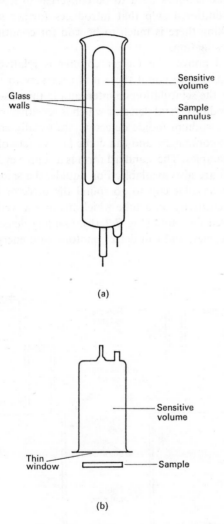

(a)

(b)

FIG. 7.1. Geiger counters for (a) liquid and (b) solid samples.

work in the field covered by this book there is the added disadvantage that wet oxidized samples need to be converted to a solid form before counting, an additional step that introduces further sources of error. Generally speaking there is much to be said for counting in solution in this kind of investigation.

As mentioned above, the Geiger counter is relatively insensitive to gamma photons, and to count these it is necessary to turn to the other type of counter, the scintillation counter, and in this case generally to one with a sodium iodide crystal detector which is sensitive to gamma photons and X-rays. The sodium iodide detectors are usually contained in light-tight aluminium containers, and can come in a variety of shapes to count solid or liquid samples. The standard form is a simple cylindrical detector, but other forms are also available. For liquids, the sample can either be presented in an annular cup to go round the detector (Fig. 7.2, (a) or, for maximum sensitivity, in a tube which fits into a well in the centre of specially prepared detectors (Fig. 7.2, (b). Gamma detectors of this kind have high efficiencies, and can detect photons with energies down to less than 0·1 MeV.

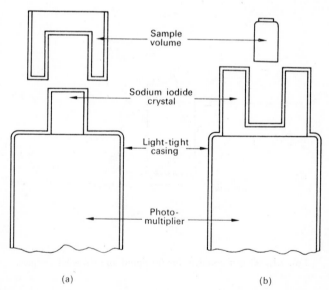

Fig. 7.2.   Scintillation counters for counting gamma emitters in solution.

For some types of nuclide neither of the types of detector so far described is suitable, because the radiation emitted is so weak that it cannot get into the sensitive volume of either of the detectors. This is true both for weak beta particles, and weak photons, and for these a special type of scintillation counter, called a liquid scintillation counter, is used. The special feature of this counter is that the detector is a complex organic chemical, and the detector and the sample are brought into close proximity by dissolving both in the same solvent. This type of counter offers high sensitivity and high efficiency, but has the disadvantage that, with the sample and detector in the same solvent, they can interfere with each other, and vary the counting efficiency. They are also more expensive than the other kinds described above.

The types of radiation which each type of counter will detect are summarized in Table 7.1.

TABLE 7.1. SUITABILITY OF DIFFERENT COUNTERS FOR DIFFERENT TYPES OF RADIATION

Type of counter	Type of radiation			
	Low energy β*	High energy β*	γ-rays	X-rays
Liquid Geiger		+		
Solid Geiger	+	+		
NaI scintillation			+	+
Liquid scintillation	+	+	+	+

* β particle denotes positively or negatively charged electrons, however arising.

The type of radiation emitted by a nuclide depends upon its mode of decay, which, for nuclides likely to be of interest in studies relevent to this book, can be one of three kinds.

The simplest decay is that in which an unstable nuclide emits a particle, to give a stable product nuclide in its ground state. This type of decay causes the emission only of particles, as is the case with say calcium 45.

The next type of decay is again by emission of a beta particle, but in this instance the product nuclide formed is in an excited state, and loses energy by emission of gamma photons. The decay is therefore characterized by the emission of both beta particles and gamma photons, as is the case, for example, with cobalt 60.

The final type of decay of interest is by electron capture. The nucleus does not, in this case, emit a particle of any kind, but instead captures one of the nuclides orbital electrons. This results in the emission of X-rays, generally of low energy, and of extra nuclear electrons, also of low energy. Generally speaking the energies of these X-rays and electrons are lower for elements of low atomic number, and higher for elements of higher atomic number. In some cases the product nuclide is again in an excited state, and loses energy by the emission of gamma-rays.

Taking the three types of decay into account, we can see that several different kinds of emission can result.

Very low energy electrons.

Low energy electrons.

High energy electrons.

Low energy X-rays.

High energy X-rays.

Gamma-rays.

Often there are several kinds of emission from one nuclide, and a choice of detectors is available but in other cases the choice is very limited. For pure electron capture decay, occurring in nuclides of fairly low atomic number, the only feasible method of counting is by liquid scintillation counting, whereas in electron capture decay of nuclides of higher atomic weight the X-rays produced are energetic enough to be counted with a sodium iodide detector.

It is now possible to look in more detail at the factors governing the choice of a suitable tracer for these studies. Obviously, if only one kind of counting equipment is available, the nuclide chosen must emit radiation compatible with it, but there are a number of other characteristics which need to be considered also.

As an example we can consider work to be done on the recovery of caesium after dry ashing. The *Radiochemical Manual* shows that there are five caesium isotopes, listed in Table 7.2.

As it is a dry ashing investigation it is desirable to be able to determine the caesium retained by the ashing vessel, and to do this requires a tracer with reasonably energetic gamma emission. This will eliminate caesium 131 which decays by pure electron capture, with the emission of relatively weak X-rays of only 0.03 MeV, which would be seriously attenuated by the material of crucibles, etc.

TABLE 7.2. CAESIUM ISOTOPES AND THEIR PROPERTIES

Isotope	Half-life	Type of decay	Particles emitted MeV	γ- and X-rays emitted MeV
Cs 131	9·7 days	E.C.		X 0·03
Cs 132	6·48 days	E.C. and β^-		γ 0·67 and others
Cs 134m	2·9 hr	I.T.		γ 0·13
				X 0·03
Cs 134	2·1 yr	β^-	0·65	γ 0·61
				γ 0·80
Cs 137	30 yr	β^-	0.51	γ 0·66 (from Ba 137m)

A second consideration is that some experiments will last at least overnight, and it is inconvenient if the tracer decays too much in that time, so that the total amount of activity decreases significantly. This factor is controlled by the half-life, which is the time period during which the amount of activity falls to half its original value, and it is clear that a half-life of 2·9 hr, for caesium 134m, is impossibly short, and that even the 6·48 days of caesium 132 is inconvenient if a series of experiments is envisaged. This process of elimination has left only two remaining nuclides, caesium 134 and caesium 137, and it might be thought that they are equally suitable. The table shows that caesium 137 decays to give a radioactive daughter nuclide barium 137m, which is the nuclide which actually emits the gamma photons. This means that if the caesium and the barium should become separated during the experiment a measurement of the gamma emission may not give a true indication of the amount of caesium present. However it must be noted that the half life of the daughter is only 2·6 min, and as it only requires about ten half-lives for parent and daughter to come to equilibrium, half an hour's delay before counting would remove the difficulty. As the two possible tracers are of very similar value technically, a final choice might well rest on commercial considerations such as price and availability.

To summarize the important points to be considered when selecting a tracer:

1. The half-life must be sufficiently long. Hopefully this would be at least several days, and preferably weeks or months. Work has been done

with nuclides having half-lives of only 10 or 12 hr, but new supplies are needed for each experiment.

2. The best type of radiation emission is gamma-rays or fairly energetic X-rays. Failing this, energetic betas, and worst of all very weak betas, or pure electron capture nuclides of low atomic number, which require liquid scintillation counting.

3. The tracer should not have a long-lived radioactive daughter which contributes significantly to the radiation emission. Strontium 90 is a poor tracer for strontium as it is the daughter, yttrium 90 that provides the most easily detectable beta-radiation. If they become separated during an experiment it is the yttrium which will be most readily traced, and due to its 2·9 day half-life, re-establishment of equilibrium will take nearly a month, and it will not be until this period has elapsed that the beta count can be taken as a measure of the strontium 90.

4. The tracer must have a sufficiently high specific activity. This is the amount of activity per gram of the element, and for trace level studies specific activities of around one curie per gram or higher are desirable, although this depends to a considerable extent on the efficiency of the counting equipment. This information is available in the radiochemical suppliers catalogues.

5. Cost. Generally speaking, reactor produced nuclides are cheaper than those produced in a cyclotron, and if other factors are more or less equal this can easily decide the choice.

After having set out the desirable features it must be recognized that in many cases there is no real choice. Some important elements effectively have no nuclide that is really suitable as a tracer, as with aluminium, while others have only a single radioactive form that may be quite satisfactory, such as beryllium 7, or very unsatisfactory, as with nickel 63. Yet again, a choice may exist, but between alternatives which are each less than ideal. This is particularly the case with the isotopes of lead, lead 210 and lead 212. Finally, there are some elements where there are two or more suitable nuclides, such as the case of caesium discussed earlier, or cobalt, and the choice can be made on other grounds.

It can be seen that the choice of tracer is not necessarily simple, and if possible, it is useful for the beginner to obtain advice. If this is not readily available from colleagues or acquaintances it is worth contacting the suppliers of radiochemicals, who have vast experience, and are ready to provide help of this kind as a service.

CHAPTER 8

INDIVIDUAL ELEMENTS

A. Lithium, Sodium, Potassium, Rubidium and Caesium

Determination of the alkali metals in various organic matrices is important in a number of fields of which the most important is probably that of human physiology.

Generally speaking these determinations are required at relatively high concentrations: demands for measurement at the part per million level are rare, except perhaps for caesium. Because the alkali metals occur widely in ionic form, it is often possible to separate them without the necessity of destroying the organic matrix first. At the simplest, many of their organic salts are water soluble and simple solution is an adequate pretreatment. For more complex substances more complicated extractions may be required, with multiple treatments being required to ensure quantitative removal of the metals.

When destruction of the organic material is required, the simplest procedure is ignition of the sample in the presence of air, and the most commonly used variant of this procedure is the use of sulphuric acid as an ashing aid. The conversion of the alkali metals to their sulphates in this way reduces the risk of loss by volatilization as these are among the least volatile of the common salts (see Table 8.1).

TABLE 8.1. MELTING POINTS OF SOME ALKALI METAL SALTS

	Melting point °C		
	Chloride	Nitrate	Sulphate
Lithium	613	255	860
Sodium	801	307	884
Potassium	776	334	1076
Rubidium	715	310	1060
Caesium	646	414	1010

The usual procedure is to wet the sample with a dilute aqueous solution of sulphuric acid, dry it gently, and then ash it, but a rather more spectacular procedure has been described[1] in which alcoholic sulphuric acid is added, and then removed by burning before the sample is transferred to the furnace.

Although the sulphated ashing method is widely quoted in standard text books, reports of difficulties can be found in the literature. Two papers published in 1931[2,3] reported low results when the sodium salts of some barbituric acid derivatives were ashed using this technique. In each case the authors found that preliminary acidification, and extraction of the organic matter, followed by sulphated ashing of the residue gave the correct result.

Similarly another worker[4] found that sulphated ashing, at an unspecified temperature, gave unsatisfactory results for sodium and potassium, and accordingly changed to a wet ashing procedure, but in the main it has proved a reliable technique for these two elements.

The ashing temperatures used will obviously be important for sulphated ashing, just as for any other kind of dry oxidation and the figures listed in Table 8.1 indicate that generally speaking, none of the alkali metals should be lost by volatilization during sulphated ashing at temperatures of 500°C or below. Sodium and potassium have regularly been treated at temperatures up to 750°C, while caesium, after conversion to sulphate, has been shown[5] to suffer virtually no loss after 48 hr heating at temperatures below 600°C (Fig. 8.1). Although a period of 48 hr heating is excessive, it is important with caesium and with rubidium to continue heating until the bisulphates and pyrosulphates first formed are converted to the sulphates.[6]

Direct ashing without an ashing aid is also a very popular procedure for alkali metals, but in this case the losses experienced will be, to some extent, a function both of the chemical form in which the element occurs, and of the nature of the material in which it is found. For biological material the

[1] Johnson, C. M., and Ulrich, A., *Bull. Div. Agric. Sciences, Univ. of Calif.* **48**, 766.

[2] Collins, G. W., *Ind. Eng. Chem.* (*Anal. Ed.*), **3**, 291 (1931).

[3] Tabern, D. L., and Shelberg, E. F., *Ind. Eng. Chem.* (*Anal. Ed.*), **3**, 278 (1931).

[4] Thomson, K. B., and Lee, W. C., *J. Biol. Chem.* **118**, 711 (1937).

[5] Ritter, R., *Naturwiss.* **51**, 104 (1964).

[6] Grant, J., *Pregls. Quantitative Organic Micro Analysis.* 5th Engl. edn., J. and A. Churchill, London, 1951.

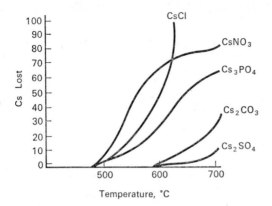

Fig. 8.1. Losses of caesium after 48hr heating (Ritter).

chloride is the commonest form, and the melting point figures given in Table 8.1 are relevant. These would seem to suggest little difficulty, except perhaps for lithium and caesium.

However, there is much evidence in the literature to suggest that the melting point data do not tell the whole story.

One extensive investigation[1] of the effect of temperature and time of heating on the recovery of sodium and potassium from a variety of animal materials showed that up to 500°C the recovery was complete even after heating for periods of 72 hr: at 550°C recoveries were still good after 24 hr heating, but after 72 hr some were low, particularly with potassium, while at 650°C nearly all the recoveries were significantly low, even at 24 hr. Further temperature increases, up to 900°C caused greater losses, but the extent varied both with the element and with the material being ashed. Less than 7% of the potassium was recovered from rabbit's blood after 24 hr at 900°C while 83% was recovered from cattle lung. The corresponding figures for sodium were 52% and 88%.

Other workers have reported conflicting results, with losses of up to 66% of potassium at 500°C on the one hand[2] and no loss from 1 mg of sodium or potassium chloride after 48 hr heating over the full heat of a Meker burner, on the other.[3] However, this last report conflicts with yet

[1] Grove E. L., Jones R. A., and Mathews, W., *Anal. Biochem.* **2**, (3), 221 (1961).
[2] Scharrer, K., and Munk, H., *Agrochimica.* **1**, 44 (1956).
[3] Kramer, B., *J. Biol. Chem.* **41**, 263 (1920).

another where only 91·5% of 1 mg of sodium chloride was recovered after a few hours at 560°C.[1]

A careful study of the volatility of some of the various chemical forms of the five alkali metals has been published[2] showing that losses at the trace level can become important at temperatures below 600°C (see Figs. 8.2 and 8.3) and that for the chlorides, with the exception of lithium chloride, the recoveries obtained bore the expected qualitative relationship to the melting points and to the vapour pressures at 550°C.

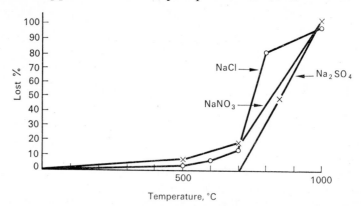

FIG. 8.2. Losses from 20 mg. of sodium salt after heating for 1hr.

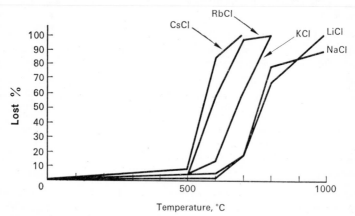

FIG. 8.3.   Losses of alkali metal chlorides on heating for 1hr (Kometani).

[1] Broadfoot, W. M., and Browning, G. M., *J.A.O.A.C.* **24**, 916 (1941).
[2] Kometani, T. Y., *Anal. Chem.* **38**, 1596 (1966).

This work showed that whereas 1g amounts of sodium chloride could be heated to 800°C for several hours with only a very small percentage loss, 20 mg amounts lost more than 70 % in 6 hr at 600°C, due to the very much larger surface to volume ratio. The results reported were obtained by heating the samples in platinum crucibles, so that the material not recovered could be assumed to have volatilized. It is important, however, when working with vessels of silica or porcelain to bear in mind the possibility of loss by reaction with the material of the dish. This type of loss is particularly probable when heating the chlorides, and the higher the temperatures used, the greater the likelihood of reaction taking place. The effect of the nature of the ashing vessel has been shown in different recoveries of potassium when ashing in silica (95 %) and porcelain (85 %).[1]

Retention losses of this kind are not easy to demonstrate using the conventional experimental techniques, and this is a field in which the use of radioactive tracers is particularly valuable, as the presence of material fixed to the surface of the ashing vessel can be demonstrated *in situ*, provided that a tracer nuclide with suitable decay characteristics can be used. The value of the tracer technique is shown in work using caesium 137 as caesium phosphate [2] where the influence of temperatures between 500°C and 700°C and of times between 0 and 48 hr on the fixation of caesium is clearly shown. By contrast, another worker, using caesium 134, demonstrated that ashing in the presence of sodium chloride gave only a 56 % recovery of caesium, but failed entirely to investigate the possibility of fixation of the tracer and attributed the loss to the formation of caesium chloride, a most improbable occurrence.

Wet oxidation methods should be uniformly successful as far as complete recovery of the alkali metals is concerned, and the choice of an individual method will rest largely upon the nature of the material to be oxidized, the levels of the elements to be determined, and the personal preference of the analyst. There is one report that Pyrex glass used in the construction of an oxidation apparatus absorbed sodium, necessitating the use of silica,[3] but generally no such difficulties have been noted.

Any of the standard wet ashing procedures described in Chapter 9 should be quite satisfactory for the alkali metals, but the possible utility of much less drastic procedures should be borne in mind. In most biological

[1] Joyet, C., *Nucleonics*, **9,** (6) 42 (1951).
[2] Ritter, R., *Naturwiss.* **51,** 104 (1964).
[3] Aubouin, G., *Radiochimica Acta*, **1,** 117 (1963).

materials and in many other organic materials, the alkali metals occur in ionic form, and provided that they can be liberated into solution, and perhaps separated from the residual matter, the degree of destruction of the residue is unimportant. With destructive techniques of determination, such as flame photometry it has sometimes been found sufficient to dissolve the whole sample and determine sodium and potassium in the resulting solution. [1] There are limitations to this approach, as most burner systems can only handle solutions low in total solids, and the method must therefore be restricted to samples containing high levels of the elements concerned, such as the homogenized fish samples described in the reference quoted. If the sample is an organic liquid such as a motor fuel, or an organic material readily soluble in a solvent this direct approach can probably be used successfully, provided that the standards are made up in a similar solvent. If the solids level is still too high, as perhaps with a lubricating grease, the organic solution can sometimes be satisfactorily stripped with an aqueous phase such as dilute hydrochloric acid which can then be used for the determination. None of these approaches can be relied upon to be uniformly successful, and with a new sample, it is always necessary to establish the suitability of the proposed method. However, for the alkali metals, some such simple technique can often prove a considerable time saver.

B. Copper, Silver and Gold

These three elements form a sub-group of low and decreasing electropositive character, forming organo-metallic compounds of low stability. Many of their compounds are readily decomposed by heating and the oxides are reduced to the metal at low temperatures either by heat alone, in the case of gold and silver, or by reducing agents, including organic matter. Many textbooks recommend the ignition of the organic compounds of these elements prior to their determination, either as the metals for gold and silver, or, for copper, as the oxide, after treatment with nitric acid to prevent low results due to reduction.

The main interest in these elements has been in the determination of small quantities in organic materials, and of the three, by far the greatest interest has been in copper. This can hardly be considered surprising in

[1] McLeod, R. A., Jones, R. E. E., and McBride, J. R., *J. Ag. Food Chem.* **8**, 132 (1960).

view of its widespread use as a material for fabrication, and its well known catalytic properties.

The determination of copper in organic materials without preliminary destruction of the organic matter has been widespread in some rather specialized circumstances. In petroleum products atomic absorption spectrophotometry has proved valuable, being applied to the product either directly, or after dilution with a suitable solvent. Similarly treatment of petroleum fractions with various acidic extractants has proved adequate to remove the copper in an ionic form.

Another field in which the ability to determine copper without first having to destroy the organic matter is of obvious importance, is in clinical chemistry, and very many methods for carrying out such determinations have been described, particularly for blood and serum. The procedure most commonly employed involves treatment of the sample with a mineral acid, generally hydrochloric acid, followed by trichloracetic acid to precipitate protein, filtration, and determination of the copper on the filtrate. Similar methods have been used for other biological liquids such as milk.

When it has proved necessary to destroy the organic material completely, as has usually been the case, a very wide range of methods has been used.

The wet oxidation methods appear to have been almost uniformly successful and nearly every possible combination has been used, ranging from sulphuric acid and potassium sulphate to sulphuric, nitric and perchloric acids plus hydrogen peroxide. Recovery experiments have been reported for a number of wet oxidation mixtures, using both chemical and radiochemical methods of determination, and some results are given in Table 8.2.

One important publication indicating less than satisfactory analytical results after destruction of organic matter by wet ashing is a collaborative report by members of the A.O.A.C.[1] in which seventeen laboratories collaborated in a study of the analysis of feedstuffs for a number of elements, including copper, by atomic absorption spectrophotometry. The coefficient of variation for copper in the samples prepared by wet oxidation with nitric and perchloric acids, was considered unsatisfactory, ranging up to 23%: unfortunately the dry ashing results were even worse.

In addition to these results of the A.O.A.C. collaborative study which

[1] Heckman, M., *J.A.O.A.C.* **50**, 45 (1967).

TABLE 8.2. RECOVERY OF COPPER AFTER WET OXIDATION

Method of oxidation*	Copper level	Recovery %	Reference
N + P + S	1— 10 ppm	95—102	1
N + P + S	14—560 ppm	101—102	2
N + P + S	10 ppm	99	3
N + P	10 ppm	100	3
N + S	10 ppm	99—101	3

* N + P + S = nitric, perchloric and sulphuric acids.
  N + P   = nitric and perchloric acids.
  N + S   = nitric and sulphuric acids.

showed dry ashing to be unreliable, there are very many reports in the literature of difficulties encountered with this sort of oxidation.

Dry oxidation has often been used with apparent success, with ashing temperatures up to 800°C being reported in the literature, and some comparisons of wet oxidations and dry ashings at temperatures of 450 to 500°C have shown little difference in the results.[4,5] On the other hand many workers have found losses occurring during dry ashing, and some of the results are given in Table 8.3.

These results consistently show low recoveries of copper after dry ashing, and when any assessment has been made of the cause of the losses the usual conclusion has been that it has occurred through interaction between the copper and the material of the crucible. Before considering this mechanism further, however, it is worth noting that there are a number of reports in the literature of errors arising through the contamination of samples with copper arising from the ashing vessel itself. This type of error has been reported for porcelain vessels on at least three occasions,[6,7,8] but its occurrence will obviously depend primarily on the composition of the glaze used, and may well vary widely from one make to another.

[1] Hubbard, D. M., and Spettel, E. C., *Anal. Chem.* **25**, 1245 (1953).
[2] Pijck, J., Hoste, J., and Gillis, J., Int. Symp. on Microchem., Birmingham, 1958.
[3] Gorsuch, T. T., *Analyst*, **84**, 135 (1959).
[4] Wyatt, P. F., *Analyst*, **78**, 656 (1953).
[5] Smit, J., and Smit, J. A., *Anal. Chim. Acta*, **8**, 274 (1953).
[6] Seitlin, H., Frodyma, M. M., and Ikeda, G., *Anal. Chem.* **30**, 1284 (1958).
[7] O'Connor, R. T., Heinzelman, D. C., and Jefferson, M. E., *J. Am. Oil. Chem. Soc.* **24**, 184 (1947).
[8] Elvehjem, C. A., and Lindow, W. C., *J. Biol. Chem.* **81**, 435 (1929).

TABLE 8.3. RECOVERIES OF COPPER AFTER DRY ASHING

Copper level	Temperature °C	Ashing aid	Recovery %	Method	Reference
10 ppm	550	none	86	Radiochemical	1
10 ppm	550	Nitric acid	94	Radiochemical	1
30 ppm	700	none	87	Radiochemical	2
30 ppm	900	none	58	Radiochemical	2
0·05 ppm	550	none	60	Chemical	3
160 mg	450	none	50	Chemical	4

Returning now to consider the mechanism of retention losses, the possibility that these might be a serious source of error was noted as early as 1934, when it was found that materials which produced only a small ash on ignition, gave erratic results, with some evidence of attack on the surface of the silica basins used.[5] This finding was supported a little later,[6] and it was also reported, that the use of highly glazed dishes reduced the losses compared with the use of old silica dishes. The effect of ashing aids was found to vary, with the bulky aids such as sodium phosphate[5] and magnesium nitrate[6] giving improved recoveries, and sulphuric acid[6] making matters worse. Such losses of copper by retention are not, of course limited to retention on apparatus, as silica present in the samples themselves is also a serious hazard, and once fixation has occurred, recovery of the copper is a difficult matter.

The normal method of solution of the ash, with hydrochloric acid, is not generally effective in removing the element from the ashing vessel, or from silica present in the sample. A mixture of acids has been claimed to be more effective,[7] and using a 2:1 mixture of dilute hydrochloric and nitric acids, good recoveries have been claimed after ashing at temperatures up to 800°C. By comparison, recoveries after extraction with hydrochloric acid alone were found to be from 15% to 60% lower. This contention that the mixture of acids is superior was not borne out by other work[8] using

[1] Gorsuch, T. T., *Analyst*, **84**, 135 (1959).
[2] Pijck, J., Hoste, J., and Gillis, J., Int. Symp. on Microchem., Birmingham, 1958.
[3] Barney, J. E., and Haight, G. P., *Anal. Chem.* **27**, 1285 (1955).
[4] Reed, J. F., and Cummings, R. W., *Ind. Eng. Chem.* **13**, 124 (1941).
[5] Tompsett, S. L., *Biochem. J.* **28**, 2088 (1934).
[6] Comrie, A. A. D., *Analyst*, **60**, 532 (1935).
[7] High, J. H., *Analyst*, **72**, 60 (1947).
[8] Gorsuch, T. T., Ph.D.Thesis, Univ. of London, 1960.

copper 64 as a radioactive tracer, in which no differences were found between the recoveries obtained with hydrochloric acid alone, and those obtained with the nitric and hydrochloric acid mixture.

The function of ashing aids in reducing losses of this type would appear to be primarily one of physical separation of the two reactive components, the dish and the copper, but a surprising finding[1] was that when small quantities of copper, as the nitrate, were heated in silica crucibles without any organic matter, or any ashing aid, that is under conditions of maximum contact, the retention of copper was very small, averaging only $0 \cdot 5\%$ in seven determinations. By contrast, when organic matter was present, and when some physical separation of the two components should have occurred, at least at first, the retention loss rose to an average of $6 \cdot 5\%$ in eight measurements. A plausible explanation of this finding is that reduction of the copper to the metal occurs in the presence of organic matter, and it is in this form that reaction with the silica takes place. This view is supported to some extent by the ready reducibility of copper compounds, and, somewhat negatively, by the fact that the copper silicates, which one would expect to be the products of reactions involving copper oxide, are unstable to acid attack.

It is difficult to draw firm conclusions from the mass of data, often conflicting, which is to be found in the literature. Wet oxidations appear to be largely successful, despite the somewhat forbidding results from the A.O.A.C. study described above, and despite a report that copper can be lost by adsorption on calcium sulphate precipitates formed during digestions involving sulphuric acid.[2] Using this type of oxidation the main criteria will again be the intractability or otherwise of the sample, and the demands of the method of determination to be applied.

The position with dry ashing procedures is less clear cut, although the weight of evidence is that the only losses of importance are those arising from fixation of the copper on some solid material. This type of interaction will be reduced by the use of low temperatures, and by the inclusion of an ashing aid. Of the many ashing aids available, magnesium nitrate has much to recommend it. It will act both as diluent and as reagent, and its use, at a temperature of about 450 to 500°C, should be satisfactory on most occasions. The use of platinum basins will probably reduce the risk of losses by reaction with the vessel during dry ashing, but will do nothing

[1] Gorsuch, T. T., *Analyst*, **84,** 135 (1959).
[2] Sylvester, M. D., and Lampitt, L. H., *Analyst*, **60,** 376 (1935).

to prevent interaction between copper and any silica present in the sample, and should not be taken as a reason for using high temperatures. Generally speaking, platinum or silica dishes should be satisfactory if used at low temperatures, and, where necessary, with an ashing aid, but there remains a slight doubt about the use of porecelain from the contamination viewpoint.

A technique which enables dry oxidations to be carried out at very low temperatures, and which is therefore of considerable interest in the determination of copper is the decomposition of organic matter with excited oxygen.[1] A recovery of copper, added to blood, of 101% is quoted as a typical result.

The recovery of silver and gold from organic matter, would be expected to be rather more difficult than the recovery of copper, particularly with dry ashing where the even greater ease of reduction to the metal should facilitate reaction with the ashing vessel. By contrast with the numerous publications on the determination of copper, the literature on the recovery of silver and gold is relatively scanty. Many textbooks recommend the ignition of organic compounds of the elements, to give the metals, prior to gravimetric determination, and platinum or porcelain crucibles are widely recommended, although attack on platinum by silver is also quoted. From the point of view of the gravimetric determination this is probably unimportant as no loss of material is caused, but for any determination involving separation of the silver from the ashing vessel, such attack would cause a serious loss. Another method widely described for dealing with compounds of gold or silver is oxidation with sulphuric acid and hydrogen peroxide.

For trace determinations both wet and dry oxidation procedures seem to have been used successfully, although sources of error can be envisaged in both types of procedure.

For silver all the usual wet ashing techniques may be expected to be successful, and radiochemical tracer experiments with silver 110 at the 10 ppm level have shown recoveries between 95 and 100% with the three usual combinations of nitric, sulphuric and perchloric acids.[2] There has been one report of difficulty in destroying silver–phthalocyanine chelates with nitric and sulphuric acids alone, although the subsequent use of perchloric acid proved successful.[3]

[1] Gleit, C. E., and Holland, W. D., *Anal. Chem.* **34**, 1454 (1962).
[2] Gorsuch, T. T., *Analyst*, **84**, 135 (1959).
[3] Cheng, K. L., *Michrochem. J.* **7**, 29 (1963).

The wet oxidation of materials containing gold could prove more difficult as heavy charring of the organic matter might lead to the separation of metallic gold with subsequent difficulties in achieving its re-solution. This type of precipitation has been reported[1] in the sulphuric and peroxide oxidation of material containing relatively large amounts of gold, and its less obvious occurrence with samples containing traces of gold might lead to errors.

The dry oxidation of materials prior to the determination of traces of silver and gold has often been described, even though it might be surmised that this would not be a satisfactory technique. Japanese workers obtained satisfactory recoveries of silver from plant materials at temperatures of 550 to 600°C, but found that at 900°C losses ranged up to 80%.[2,3] Similarly, although Sandell[4] recommends an ashing temperature below 500°C, higher temperatures, 550 and 600°C, have also been used.

A number of tracer investigations into the recovery of silver and gold after dry ashing have been carried out, which support the view that this is not a suitable technique for these elements. The results are listed in Table 8.4.

TABLE 8.4. RECOVERY OF SILVER AND GOLD AFTER DRY OXIDATION

Element	Organic material	Temperature °C	Recovery %	Reference
Silver	Cocoa	550	93, 99	5
Silver	Blood	400	65	6
Silver	Blood	700	45	6
Silver	Blood	900	21	6
Silver	Blood	< 100	72	7
Gold	Blood	400	19	6
Gold	Blood	700	0	6
Gold	Blood	900	0	6
Gold	Blood	< 100	70	7

[1] Tabern, D. L., and Shelberg. E. F., *Ind. Eng. Chem. (Anal. Ed.)*, **4**, 401 (1932).
[2] Kudo, Kiyoshi, *J. Chem. Soc. Japan. Pure Chem. Sect.* **79**, 693 (1958).
[3] Suzuki, Nobuo, *Nippon Kagaku Zasshi,* **80**, 269 (1959).
[4] Sandell, E. B., *Anal. Chem.* **20**, 253 (1948).
[5] Gorsuch, T. T., *Analyst*, **84**, 135 (1959).
[6] Pijck, J., Hoste, J., and Gillis, J., Int. Symp. on Microchem., Birmingham, 1958.
[7] Gleit, C. E., and Holland, W. D., *Anal. Chem.* **34**, 1454 (1962).

Large losses are found in nearly all instances, including, rather surprisingly the low temperature ashing procedure with excited oxygen.[1] In this case the material not recovered from within the ashing vessel was found in the trap, so that a distillation mechanism would seem to have operated, although a retention mechanism would appear more probable.

Compounds of silver and gold will be reduced to the metals by heat alone at the temperatures commonly used for dry ashing, so that in the presence of charred organic material this reduction can be expected to be complete and rapid. Consideration of the mechanism of any retention can therefore be limited to the metallic state, and a diffusion process involving non-ionic forms of the metals seems likely. Radiochemical studies[2] have shown that when gold and silver solutions are dried in silica crucibles, without organic matter, and heated at 500°C for 16 hr, very large percentages are retained by the crucibles.

Work carried out with ceramics and with silicate melts is of interest here, and the following quotation from a paper by Forland and Weyl[3] is directly relevant. "There exists a general trend, that increases from copper to the noble elements, gold and platinum, to form the metallic phase rather than a compound. Silver...is known to be in equilibrium with metallic silver in silicate glasses. With gold and platinum the equilibrium must be expected to be shifted far to the side of the metal." Although this line of argument may explain losses occurring by retention on the ashing vessel, it does not explain the apparently large volatilization losses reported, and there is clear scope here for further investigation.

One consequence of the reduction of gold compounds to the metal during ashing is the need to use a suitable solvent such as aqua regia, or hydrochloric acid and bromine, for solution of the ash.

C. Beryllium, Magnesium, Calcium, Strontium, Barium and Radium

The elements of this group, with the exception of beryllium, are among the least demanding when it comes to considering methods for the removal of organic matter. Their common salts are non-volatile and stable, (see Table 8.5) and they have no marked tendency to form volatile organic compounds or complexes. Of the five elements listed only beryllium has

[1] Gleit, C. E., and Holland, W. D., *Anal. Chem.* **34,** 1454 (1962).
[2] Gorsuch, T. T., *Analyst,* **84,** 135 (1959).
[3] Forland, T., and Weyl, W. A., *J. Amer. Ceram. Soc.* **31,** 105 (1958).

been found to be troublesome to any significant extent, and, in view of its different chemical properties, it is convenient to treat it separately.

TABLE 8.5. MELTING POINTS OF ALKALI METAL COMPOUNDS

Element	Melting point °C				
	Chloride	Nitrate	Sulphate	Oxide	Fluoride
Beryllium	440	60	550–600 d.	2500	800
Magnesium	708		1124 d.	2800	1396
Calcium	772	561	1450	2580	1360
Strontium	873	570	1580 d.	2430	1190
Barium	962	592	1580	1923	1280

1. Beryllium

The considerable interest which has arisen in the toxicological significance of beryllium, and the very low levels at which it is considered to be a hazard has led to a demand for the determination of very small quantities. Both wet and dry oxidation procedures have been used to achieve the necessary preliminary separation, and some difficulties have been experienced. Wet oxidation has generally proved satisfactory, and probably any combination of the common oxidizing reagents, which will handle the organic matter, will give a reasonable recovery of beryllium; the problem of contamination is more likely to be troublesome. On the other hand although several workers have found dry ashing quite satisfactory, and reported good recoveries, others have been much less successful, as can be seen from the results in Table 8.6.

A plausible explanation for discrepancies of this type has been indicated by the work of Toribara and his co-workers[1,2] who showed, using beryllium 7 as a tracer, that although no beryllium was lost when 20 mg amounts were heated in platinum at 750°C for 17 hr, the material so treated would not then react with complexing agents and ion exchange resins in the expected manner. This was attributed to the formation of a very insoluble beryllium oxide, which would not take part in the typical reactions of beryllium, and so was effectively lost even though actually still present. In the light of this explanation, it is interesting to note that

[1] Toribara, T. Y., and Chen, P. S., *Anal. Chem.* **24,** 539 (1952).
[2] Toribara, T. Y., and Sherman, R. E., *Anal. Chem.* **25,** 1594 (1953).

TABLE 8.6. RECOVERY OF BERYLLIUM ON DRY ASHING

Quantity of beryllium	Recovery %	Comments	Reference
2·5mg	118		
5 mg	105		1
10—45 mg	100		
1 mg	65		
2 mg	50		
4 mg	65	Silica dish	2
4 mg	12·5	Platinum dish	
8 mg	40		
16 mg	38		

Hiser *et al.* (Table 8.6.), who obtained good recoveries of beryllium, treated the residue from the ashing with hydrofluoric acid to take it in to solution.

2. Magnesium, Calcium, Strontium and Barium

The individual members of this group of elements are of varying importance in a number of fields, particularly in the various aspects of the biological sciences. Magnesium and calcium are major constituents of many animal tissues, while barium, and to some extent strontium, are poisons. In recent years the incorporation of strontium and barium into plant and animal tissues has assumed further importance due to their presence in high abundance in the products of nuclear fission.

As with the alkali metals, a number of procedures have been described in which complete destruction of the organic matter is not required. Generally these apply to samples that are already liquid, such as blood, urine or petroleum fractions, but this is not always the case.

The simplest treatment is solution of the sample in an appropriate solvent, followed by the application of a suitable method of determination. This approach has proved particularly popular as a preliminary to the determination of calcium and magnesium by flame photometry or by atomic absorption spectrometry. Urine samples generally need nothing more than dilution with water, dilute acid, or a simple salt solution, before aspiration into the flame, but blood fractions often need deproteinizing first. Petroleum fractions, too, often require no more than dilution before

[1] Hiser, R. A., Donaldson, H. M., and Schwenzfeier, C. W., *Ind. Hyg. J.* **22**, 280 (1962).

[2] Cholak, J., and Hubbard, D. M., *Anal. Chem.* **20**, 73 (1948).

measurement, and a system has been described,[1] in which the dilution and the determination of several elements, including magnesium, is carried out automatically. However, in this automatic procedure, although the recovery of known standards showed an average recovery of 100% for magnesium, a comparison with the results obtained on some of the samples after dry ashing showed a tendency for the results of the automatic procedure to be lower.

Titration with EDTA has also been employed for determination of calcium and magnesium, either on solutions of the samples, or even on the sample itself. Titration of calcium in urine and serum has been particularly studied, and a host of different indicators have been used for the purpose.

Slightly more complicated procedures have been described involving extraction of the sample with various reagents, with acetic acid–sodium acetate buffers being popular choice. Extraction with EDTA solutions has given results for magnesium and calcium which were considered to agree well with dry ashing at 550°C[2] while evaporation of up to 1 g of sample with 5 ml of concentrated nitric acid has been found to be sufficient pretreatment of samples including oyster shells, flesh and vegetable matter before determination of calcium by X-ray fluorescence.[3]

The complete oxidation of organic materials should cause few problems for the recovery of this group of elements. For organo-metallic compounds which contain no other involatile components simple ashing is widely recommended in the text books, generally without an ashing aid for magnesium, but by sulphated ashing for calcium, strontium and barium. The element is then determined by weighing. A similar ashing procedure can readily be used for organic and biological material containing only small quantities of the elements, although here it is probably wise to avoid the use of sulphuric acid, as it will be necessary to redissolve the residue for further treatment, and with barium particularly, the presence of sulphate is likely to cause trouble. The choice of ashing temperature is not critical, and temperatures between 500 and 600°C are probably quite satisfactory, with 550°C a reasonable compromise. Much higher temperatures have been used, but one worker has reported losses of magnesium of 7% on ashing grass samples at 740°C, whereas temper-

[1] Slavin, S., and Slavin, W., *Atomic Absorption Newsletter*, **5**, 106 (1966). Published by Perkin Elmer, Norwalk, Conn., U.S.A.

[2] Greweling, T., *J. Ag. and Food Chem.* **10**, 138 (1962).

[3] Campion, K. P., and Whittem, R. N., *Analyst*, **92**, 112 (1967).

atures of 430, 530 and 620°C caused no losses.[1] The amounts of magnesium in the samples were in the 1–10 mg region, so the loss of 7% could represent a considerable amount of the element.

Losses of magnesium have been reported at very much lower temperatures, and a relatively recent paper[2] has put the ashing temperature limit at 450°C but this does not seem to accord with general experience. The A.O.A.C. Methods of Analysis quotes temperatues between 500 and 550°C.

Calcium, too, has been said to be lost at very low temperatures, with a loss of 10% at 400°C being reported,[3] but there is a bigger body of evidence that much higher temperatures are quite satisfactory. A collaborative study[4] on the analysis of poultry feed and dairy feed showed no significant differences in the results obtained after ashing at 500, 600 and 700°C and all the results were a little higher than those obtained by the wet ashing referee method. Similarly a later study[5] on petroleum oils, using calcium naphthenate, labelled with calcium 45, showed no significant losses when oils containing 6 to 144 ppm of calcium were subjected to controlled burning and ashing at 550°C, while another A.O.A.C. collaborative study found no significant difference in the recoveries of calcium and magnesium from feeds after dry ashing at 550°C for 4 hr, and after wet oxidation.[6]

Investigations into the recovery of strontium and barium have been fewer, although Thiers[7] obtained good recoveries of both at the 1 ppm level after careful ashing at 450 to 500°C, and there seems to be no reason to suppose that there will be any difficulties at the temperatures used for calcium.

A rather different type of dry ashing which has been used for small samples (10 to 20 mg) before the determination of these elements, is the oxygen flask method, and no particular difficulties have been reported. For some applications, where the samples are small, the speed of the method is very convenient, but its use is very much the exception for normal work.

[1] Davidson, J., *Analyst*, **77**, 263 (1952).
[2] Fawcett, J. K., and Wynn, V. J., *Clin. Path.* **14**, 403 (1961).
[3] Griggs, M., Johnstin, R., and Elledge, B., *Ind. Eng. Chem. (Anal. Ed.)*, **13**, 99 (1941).
[4] St. John, J. L., *J.A.O.A.C.* **24**, 848 (1941).
[5] Morgan, L. O., and Turner, S. E., *Anal. Chem.* **23**, 1365 (1951).
[6] Heckman, M., *J.A.O.A.C.* **50**, 45 (1967).
[7] Thiers, R. E., in Glick, D., Ed., *Methods of Biochemical Analysis*, Vol. 5, p. 273. Interscience, New York, 1957.

Wet oxidation methods have been widely used for all of these elements, and in general little trouble has been found except perhaps by the workers who made the rather gloomy comment that "In particular the wet ashing technique . . . is not only exceedingly laborious and unpleasant, but is dangerous, both to the operator and to the sample." The only necessary word of caution concerns the use of sulphuric acid in the various oxidation mixtures. For magnesium there is no problem, and for calcium in small amounts the solubility of the sulphate is usually sufficient, but for samples containing large amounts of calcium, and for strontium and barium at all times, it is probably wise to avoid the use of sulphuric acid altogether. It is, however, interesting to note that in one tracer study, using 10 ppm of strontium labelled with strontium 89, recoveries between 96 and 100% were achieved after oxidation with mixtures containing sulphuric acid. [1] The recoveries were determined by radioactivity measurements on the solutions remaining after completion of the oxidation, and the possibility must be considered that the strontium was present as a finely divided suspension of strontium sulphate, which would not necessarily be available for determination by any other means.

The solubility of the alkaline earth sulphates are given in Table 8.7.

TABLE 8.7. SOLUBILITIES OF THE ALKALINE EARTH SULPHATES IN COLD WATER

Element	Solubility g/100 cc
Magnesium	26 @ 0°C
Calcium	0·21 @ 30°C
Strontium	0·011 @ 0°C
Barium	0·00022 @ 18°C
Radium	0·000002 @ 25°C

The disadvantage of being unable to use sulphuric acid lies in the extent to which the rate of oxidation of the sample will thereby be decreased, because of the lower temperatures reached. This can be overcome by using a mixture of perchloric and nitric acids, an oxidation system which has many advantages, and which has been recommended as the best all purpose mixture. [2] The possible hazards, and the precautions necessary, when using perchloric acid have been discussed earlier. A mixture of nitric

[1] Gorsuch, T. T., *Analyst*, **84**, 135 (1959).
[2] Nangiot, P., *J. Electroanalyt. Chem.* **7**, 50 (1964).

acid and hydrogen peroxide can also be used, and the availability of the reagent in concentrations of 50% or more and in a high state of purity will probably extend the use of this mixture.

The separation of traces of radium from organic samples probably presents few additional problems over and above those found with barium, except that the very low solubility of its sulphate argues even more strongly against any possible use of sulphuric acid.

D. Zinc, Cadmium and Mercury

These three elements form a sub-group of increasing density, increasing atomic volume, increasing volatility of the metals, and increasing problems of recovery from organic materials.

1. Zinc

Although zinc has often been cited as a material difficult to recover quantitatively from organic matrices, it does, in fact, present relatively few problems.

The determination of zinc in some of the simpler types of organic systems, such as oil additives, blood serum, urine, or some pharmaceutical compounds can frequently be carried out after extraction of the sample rather than destruction of the organic matter, while techniques such as X-rays fluorescence or atomic absorption can sometimes be applied with even less preliminary treatment.

More vigorous methods (including both wet and dry ashing techniques) have been applied to many zinc-containing organic compounds with apparent success. Dry ashing alone or with sulphuric acid is commonly recommended in textbooks for compounds which contain no involatile material except zinc, while Carius oxidation and oxygen flask combustion are other techniques well suited to the decomposition of organo-zinc compounds.

Turning to consideration of smaller amounts of zinc in organic matrices, most of the common oxidation procedures have been applied, but whereas wet digestions seem to have given uniformly good recoveries, dry ashing methods have aroused an appreciable amount of controversy.

Radio tracer techniques have been used to investigate briefly the major wet oxidation procedures, and some recovery figures are given in Table 8.8.

TABLE 8.8. RECOVERY OF ZINC AFTER WET OXIDATION

Oxidation method	Recovery %	Reference
Nitric and perchloric acids	98, 100	1
Nitric, perchloric and sulphuric acids	101, 94	1
Nitric and sulphuric acids	101, 99	1
Nitric, perchloric and sulphuric acids	99, 102	2
Sulphuric acid and hydrogen peroxide	100, 102	3

Dry ashing techniques have also been very popular, but the conditions used and the results obtained have varied widely. Ashing temperatures at least up to 900°C have been reported in the literature, although most values have been much lower, while recoveries have ranged as low as thirty per cent. A selection of the results of investigations into losses arising during dry oxidation procedures is given in Table 8.9.

TABLE 8.9. RECOVERY OF ZINC AFTER DRY ASHING

Temperature °C	Ash-aid if any	Ashing vessel	Method of investigation	Recovery %	Reference
550		Silica	Radiochemical	96	
550	HNO$_3$	Silica	Radiochemical	97	
550	H$_2$SO$_4$	Silica	Radiochemical	100	1
550	Mg(NO$_3$)$_2$	Silica	Radiochemical	99	
450			Chemical	70–90	4
450		Silica	Radiochemical	~100	
550		Silica	Radiochemical	95	5
850		Silica	Radiochemical	55	
500		Silica	Radiochemical	77	
700		Silica	Radiochemical	36	
900		Silica	Radiochemical	11	6
800		Platinum	Radiochemical	~100	
500			Chemical	50–93	7
500		Silica	Radiochemical	98	
700		Silica	Radiochemical	69	8
900		Silica	Radiocehmical	30	

[1] Gorsuch, T. T., *Analyst*, **84**, 135 (1959).
[2] Pijck, J., Hoste, J., and Gillis, J., Int. Symp. on Microchem., Birmingham, 1958.
[3] Down, J. L., and Gorsuch, T. T., *Analyst*, **92**, 398 (1967).
[4] Nangiot, P., *J. Electroanalyt. Chem.* **7**, 50 (1964).
[5] Hamilton, E. I., Minski, M. J., and Cleary, J., *Analyst*, **92**, 257 (1967).
[6] Spitzy, H., and Dosudil, I., *Mickrochimica Acta*, 119 (1962).
[7] Scharrer, K., and Munk, H., *Agrochimica*, **1**, 44 (1956).
[8] Pijck, J., Hoste, J., and Gillis, J., Int. Symp. on Microchem., Birmingham, 1958.

The evidence of these results is conflicting, with losses in some cases being reported after ashing at temperatures of 500°C and below.

The two types of loss possible in a dry ashing situation such as this, excluding purely mechanical losses, are losses by volatilization, and losses by retention, and both types have been postulated to explain low zinc recoveries.

Volatilization losses have been explained both by the formation of zinc metal by reduction of zinc oxide by carbon, and by the formation of zinc chloride by interaction of zinc salts with sodium chloride, or other chlorine containing materials. Zinc metal melts at 419°C and boils at 907°C while zinc chloride melts as low as 262°C, so that losses would be expected at normal ashing temperatures, if either of them were formed (see Table 8.10).

TABLE 8.10. TEMPERATURES (°C) AT WHICH THE SUBSTANCES LISTED HAVE THE VAPOUR PRESSURE SHOWN

	1 mm Hg	10 mm Hg
Zinc	487	
Zinc chloride	428	
Cadmium	394	
Mercury	126	184
Mercuric chloride	136	180

However, in the case of the formation of the metal by reduction with carbon, the free energy data[1] show that at all temperatures below about 950°C the free energy changes for the reactions

$$2Zn + O_2 = 2ZnO$$

and $$2C + O_2 = 2CO$$

are such that the reaction

$$ZnO + C = Zn + CO$$

is not feasible. It therefore seems safe to eliminate this mechanism as a possible explanation for losses of zinc.

[1] Ellingham, H. J. T., *J. Soc. Chem. Ind.* **63**, 125 (1944).

In a similar manner it can be shown that the free energy change at 600°C ($\triangle G_0$) for the reaction

$$ZnO + 2NaCl = ZnCl_2 + Na_2O$$

is $+ 79$ kcal, so that this reaction is again not feasible thermodynamically. However, $\triangle G_0$ for the reaction

$$ZnO + 2HCl = ZnCl_2 + H_2O$$

is -11 kcal at 600°C, so this reaction can be expected to proceed.

In support of these contentions recovery experiments have been carried out in which zinc nitrate was heated with small quantities of sodium chloride or ammonium chloride (which can be expected to dissociate to give hydrochloric acid at high temperatures). The results are given in Table 8.11.

TABLE 8.11. RECOVERIES OF ZINC AFTER HEATING WITH INORGANIC CHLORIDES

Chloride added	Recovery %
None	100, 104
Sodium chloride	102, 100
Ammonium chloride	7, 7, 9, 6

These results indicate that any compound which is likely to produce hydrochloric acid on heating must be regarded with suspicion in this context, and this will include organo-chlorine compounds as well as inorganic compounds. However, most biological materials are not likely to cause difficulty in this way, as the form in which chlorine is most likely to occur is as sodium chloride, and this does not seem to cause volatilization losses.

Similarly Table 8.12 shows the results of an experiment in which human hair, which naturally contained a small amount of zinc, was irradiated with neutrons to produce zinc 65 *in situ*. A measure of the quantity of this zinc was obtained by gamma spectrometry, and the hair was ashed. The zinc was redetermined by gamma spectrometry, and the ash repeatedly heated and measured, using progressively higher furnace temperatures.

TABLE 8.12 BEHAVIOUR OF ZINC ON ASHING

Thermal history	Recovery	
	Sample A %	Sample B %
Intact sample	101	94
500°C — 16 hr	100	100
+ 600°C — 3 hr	100	100
+ 700°C — 3 hr	100	100
+ 800°C — 3 hr	102	95
+ 900°C — 3 hr	100	96
+ 1000°C — 3hr	97	97

There is no evidence in these figures for volatilization of zinc from the sample, even under an extremely rigorous heating programme which involved 31 hr of heating at temperatures up to 1000°C. However, when the ashes of these two samples were finally dissolved out in acid, more than one-third of the zinc was found to be firmly retained on the silica ashing vessels. This finding can be explained by the formation of a stable zinc silicate by a solid state reaction between the zinc compound and the material of the ashing vessel. It has been shown that this effect is accentuated by the presence of sodium chloride which presumably acts by weakening the silica structure and facilitating the reaction with the zinc compound, and in the experiment summarized in Table 8.11 although no zinc was lost by volatilization in the presence of sodium chloride, only some ten per cent of it could be removed from the silica crucibles used and recovered in solution. The other 90% was firmly bound to the silica.

It therefore seems probable that the presence in a sample of large amounts of chlorine in any chemical form, must be regarded with caution, as it may bring with it the risk of serious losses either by volatilization or by retention. However, given this proviso, there seems no reason why dry ashing at a temperature of 500°C should not be quite satisfactory for the the majority of materials.

2. Cadmium

Despite the fact that cadmium is quite toxic, and has been responsible for a number of cases of poisoning, relatively little work has been done in the study of its behaviour during the oxidation of organic materials.

Wet oxidation procedures have been widely favoured for the preliminary decomposition stage, and many types of procedure have been used without difficulty. The results of some tracer recovery experiments, using a variety of techniques, are given in Table 8.13.

TABLE 8.13. THE RECOVERY OF CADMIUM AFTER WET ASHING

Oxidation mixture	Material	Recovery %	Reference
Nitric and sulphuric acids	Cocoa	103, 100	
Nitric and perchloric acids	Cocoa	100, 101	1
Nitric, sulphuric and perchloric acids	Cocoa	102, 100	
Sulphuric acid and hydrogen peroxide	PVC	94, 95, 97, 97	
Sulphuric acid and hydrogen peroxide	Polythene	95, 97	2

Consideration of these and other results indicates that there should be no difficulty in recovering cadmium from virtually any of the normal types of organic sample by the common wet oxidation methods.

The situation is, however, rather different when organic material is dry ashed, prior to the determination of cadmium. Losses have been reported by a number of workers under a variety of dry ashing conditions, although a few people appear to have used the method successfully, or at least without adverse comment. Ashing temperatures as low as 500°C have been found to lead to losses of cadmium. [3] Other studies showed that recoveries were lowest when deflagration occurred, as might be expected, but also more surprisingly, that the addition of nitric acid to hasten the oxidation, made the losses even greater. [4] This finding was borne out in tracer studies in which large amounts of cadmium—up to twenty per cent of the 10 ppm of cadmium added—were found to volatilize when either nitric acid or magnesium nitrate were used as ashing aids. It is interesting to note that the same study showed no losses when the organic material was oxidized by the method described by Middleton and Stuckey [5] in which nitric

[1] Gorsuch, T. T., *Analyst*, **84**, 135 (1959).
[2] Down, J. L., and Gorsuch, T. T., *Analyst*, **92**, 398 (1967).
[3] Klein, A. K., and Wichman, H. J., *J.A.O.A.C.* **28**, 257 (1945).
[4] Cholak, J., and Hubbard, D. M., *Ind. Eng. Chem. (Anal. Ed.)* **16**, 333 (1944).
[5] Middleton, G., and Stuckey, R. E., *Analyst*, **79**, 138 (1954).

acid is used throughout, but in which the temperature does not rise much over 300°C.

It is difficult, in the absence of more detailed information, to make any firm recommendations about the use of dry ashing prior to the determination of cadmium. It is not, at present, clear why the use of nitric acid should have such a bad effect on the recovery, although it is possible to speculate as to the possible mechanisms involved.

The free energy data of Ellingham[1] indicate that cadmium oxide is likely to be reduced by carbon at a temperature of about 600°C. As cadmium metal melts at 320°C and boils at 778°C any such reduction is likely to lead to low recoveries (see Table 8.10). Paradoxically, the use of ashing aids such as nitric acid, or magnesium nitrate might enhance the loss by causing the temperature to rise locally to levels markedly above the nominal furnace temperature so that this type of reduction might occur in regions of the sample where reducing conditions still exist.

It is clear that further experimental work will be required to support or demolish this hypothesis, but it offers an explanation of the facts already observed.

3. Mercury

Following on from zinc and cadmium, we come to an element which is liquid at room temperature, boils at 360°C, has compounds which are, in the main, either volatile or readily decomposed, and which has probably caused more difficulties, in the context of the procedures under consideration in this monograph, than any other.

Partly as a consequence of the special properties of mercury, there is probably a greater divergence between the methods used for the organo-metallic compounds, and those used for trace determinations, than there is with most other elements.

Because of the difficulties involved in any operation in which mercury compounds are heated to high temperatures, many procedures have been described in which the disruption of the sample stops far short of complete destruction of the organic moiety. One popular group of methods, of varying severity, involves treatment of the organo-metallic compound with hydrochloric acid. The treatement used has varied from a 30 sec shaking, to some hours refluxing, with, where necessary, solution of the material in a suitable solvent to facilitate the separation.

[1] Ellingham, H. J. T., *J. Soc. Chem. Ind.* **63**, 125 (1944).

Complexing agents, other than chloride ion that have also been used for the removal of mercury from organic combination, include iodine in acid potassium iodide solution, thiocyanate ion, and sulphide solution.

Another widely used type of determination involves the reduction of mercury compounds to the metal, followed, in some cases, by amalgamation with zinc. The reduction is often carried out by heating the sample, usually of the R—Hg$^+$ type, with zinc and an acid, and the acids used have included formic, acetic and hydrochloric, although presumably many others would also be suitable. In a related procedure the reduction is carried out with stannous chloride and hydrochloric acid, and this is said to give a method of more general application.

These relatively mild treatments are generally not suitable for all types of mercury compounds, and more vigorous treatments have also been widely applied.

The standard type of combustion train oxidation can be applied to mercury compounds, but in this instance, instead of the metal remaining behind as a residue, as occurs with most elements, it distils over and is trapped and measured, commonly by using a gold foil as a collector. This type of combustion uses air, or oxygen as the flowing medium, but comparable methods have been described in which the sample is pyrolysed and reduced in a stream of hydrogen, although such procedures have never become widely accepted.

Oxygen flask combustions have been described on many occasions, and the suitability of this technique for the recovery of a metal as volatile as mercury, is clear. Similar arguments can be applied to Carius type oxidations using nitric acid in sealed tubes, and this method has also been widely recommended.

The use of vigorous oxidation procedures in open vessels has been described for organo-mercury compounds, and the methods used have presumably been successful. Nitric acid, sulphuric acid and sodium persulphate, and sulphuric acid and hydrogen peroxide, have been used, among others, for these determinations, but whereas some workers specify the use of air or water condensers, others, using similar methods, have not apparently found them necessary.

When determining traces of mercury in organic or biological materials there can be little doubt that the major problem lies in destroying the organic matter and converting the mercury into a simple inorganic form in solution. So difficult has this problem proved to be that many attempts have been made, as with the organo-mercury compounds, to circumvent it, by settling for much less than complete destruction. In some instances, generally involving samples such as urine, which contain relatively little organic matter, this has clearly been successful, but this type of procedure has often been criticized because of the risk that the partially degraded organic matter may still be able to form strong complexes with the mercury, and so make it difficult or impossible to measure. Although there is something satisfying to the analytical soul in transforming a mercury-containing organic sample into a simple inorganic solution containing all the mercury and none of the organic matter, it is clearly unwise to strive to obtain the latter at the risk of failing in the former, if a satisfactory determination can be carried out on material less comprehensively destroyed.

The successful use of less than total oxidation has often been reported with mixtures containing an acid—generally sulphuric acid—and potassium permanganate. The prime need for the satisfactory recovery of mercury is said to be the destruction of thiol groups which form very strong mercury linkages, and this can readily be achieved with permanganate mixtures, although the severity of the treatment applied has varied widely. Perhaps the mildest treatment of this kind is the one in which small amounts of blood or tissue were treated with the oxidation mixture at 0°C and briefly warmed to 50°C, although this cannot be called a typical application. More widely used and rather more severe, is the procedure described for urine[1] and since adopted as a standard for the analysis of trade effluents,[2] in which the sample is heated at 85°C in a sealed bottle. Still more vigorous is the treatment first applied to urine in the early years of the century, in which the sample is refluxed with sulphuric acid and permanganate. This is still widely applied to urine and has also been extended to the oxidation of faeces or tissue after homogenization.[3]

As the samples to be analysed become more solid, the methods used for the preliminary disruption of the sample tend to move nearer to the goal of the complete destruction of the organic matter.

Dry ashing is seldom used for this purpose, except in the rather rare procedures in which the volatility of mercury is changed from a liability to an asset, by distilling the metal from the sample before determination. Such methods seems to have notable advantages compared with the normal

[1] Rolfe, A. C., Russell, F. R., and Wilkinson, N. T., *Analyst*, **80**, 523 (1955).
[2] A.B.C.M.—S.A.C. Committee, *Analyst*, **81**, 176 (1956).
[3] Gage, J. C., *Brit. J. Ind. Med.* **18**, 287 (1961).

oxidation methods, but their use is far from common. This is, perhaps, due to the fact that appreciable quantities of volatile organic matter distil over with the mercury, so that the problem of removal is deferred and simplified, rather than solved.

One method of this kind, applied to organo-mercury compounds rather than to samples containing traces of mercury, involved combustion at 750 to 800°C, and passage of the combustion products through a subsidiary furnace containing a suitable catalyst, before absorbing the mercury in a tube of silver granules.[1] This fairly complex procedure was necessary for samples where the mercury to organic matter ratio was high, so it is easy to understand the much greater problems to be faced when the ratio is vastly less favourable.

Wet ashing techniques are far and away the most commonly used methods for preparing this type of organic sample for the determination of mercury, and nitric–sulphuric mixtures are easily the most widely used. The official A.O.A.C. method, based on the work of Klein, is of this type and involves digestion of the sample in a special apparatus which allows part of the sample to be distilled into a reservoir and reserved. This allows the temperature of the acid mixture remaining in the flask to rise, so that the organic matter is progressively destroyed. This procedure will not destroy fat, which must be filtered off, but there is no evidence that inorganic mercury is lost because of this. This method has been widely used, but tracer studies made in the course of an investigation into the determination of mercury[2] indicated that, at least with mercury levels below 1 ppm substantial amounts of mercury, averaging about 15%, were transferred to the reservoir with the nitric acid distillate. To overcome this difficulty a method was evolved in which the distillate and the final digest were recombined prior to the determination, and it was shown that, despite the presence of small organic molecules in the distillate, there was no interference with the final determination.[2] This method is probably the best available today for general use before the determination of mercury.

Various specialist procedures have been described, which make use of nitric–sulphuric oxidation. One such is the method for dealing with large samples of apples or tomatoes, using selenium powder to "fix" the

[1] Mitsui, Tetsuo, Yoshikawa, Keikichi, and Sakai, Yosuke, *Michrochem. J.* 7 160 (1963).
[2] Analytical Methods Committee, *Analyst,* **90,** 515 (1965).

mercury, and prevent its loss. This method has been adopted as an official method for the analysis of these materials[1] and has also been applied, in a slightly modified form, to the analysis of urine and faeces, giving recoveries between 88 and 98%.[2] It is interesting to note that an earlier investigation into the determination of mercury on the peel of apples found poor recoveries with fresh material, but satisfactory results on the freeze dried material.[3] This was attributed to dilution of the oxidation mixture by the water in the sample, a source of error that could easily be eliminated by use of the Analytical Methods Committee (1965) method, referred to on p. 82 and described on pp. 140 to 141.

The use of oxidation mixtures containing perchloric acid has also been investigated in some detail, using radio tracer techniques, and a number of problems peculiar to these methods discovered.[4] The obvious advantage of such procedures, quite apart from the significant gain in oxidation potential, is that a temperature ceiling of about 200°C is very readily maintained, so long as perchloric acid is present. It might reasonably be expected that this relatively low temperature would minimize the problems of mercury recovery, but instead it was found that losses were high and that they varied markedly with the type of organic material being destroyed. These losses appeared to be due to the high temperature reduction of perchloric acid to hydrochloric acid, and volatilization of mercuric chloride. A method was devised to overcome this problem by trapping the distillate and redistilling it with sulphuric acid, but, although it works well, it is rather less convenient than the A.M.C. method quoted above. More recently[5] a method has been described in which oxidation with nitric and perchloric acid is carried out in a flask fitted with a tube partly packed with Raschig rings. No indication of the effectiveness of this procedure was given.

The use of sulphuric acid and hydrogen peroxide has not very often been described for the destruction of organic matter prior to the determination of mercury, despite the advantages to be gained from the use of this mixture. One paper[6] describes a procedure in which the sample is slurried with concentrated sulphuric acid, and 50% hydrogen peroxide added drop

[1] Analytical Methods Committee, *Analyst*, **86**, 608 (1961).
[2] Tompsett, S. L., and Smith, D. C., *J. Clin. Pathol.* **12**, 219 (1959).
[3] Arthington, W., and Hulme, A. C., *Analyst*, **76**, 211 (1951).
[4] Gorsuch, T. T., *Analyst*, **84**, 135 (1959).
[5] Epps, E. A., *J.A.O.A.C.* **49** (4), 793 (1966).
[6] Polley, D., and Miller, V. L., *Anal. Chem.* **27**, 1162 (1955).

by drop, first in the cold and then with warming, but the application of what might be regarded as the normal technique with these reagents has not been reported.

Although the usual types of dry ashing procedures are of little relevance in this discussion, oxidation of relatively large samples with oxygen has been carried out successfully using a modified oxygen flask procedure[1] with a 5-litre flask. Approximately 2g quantities of dried apple were burned in a special flask, fitted with an expansion device to allow for pressure changes.

The low temperature ashing technique of Gleit and Holland, using excited oxygen has also been used in mercury recovery experiments, but without conspicuous success. [2]

All attempts to determine small amounts of mercury in organic samples are bedevilled by the problems caused by the volatility of the metal and its compounds, which make it very difficult to recover the element quantitatively. One solution to this would be to include a recovery indicator in each determination, so that however erratic the recoveries, accurate corrections could be made for them. Such corrections could, in fact, be applied quite readily by the use of radioactive mercury 203 which could be added with the acid oxidation mixture at the beginning, and determined by gamma counting on the final solution. The amount of activity required for such a procedure need only be about $0 \cdot 05$ μc, and at a specific activity of one curie per gram, this would only be associated with $0 \cdot 05$ μg of chemical mercury, an amount that could be ignored in the majority of determinations.

This technique can obviously be applied also to other elements, but the long and troubled history of mercury determinations makes its application to this element particularly apposite.

E. Aluminium, Gallium, Indium and Thallium

Relatively little work has been carried out on the determination of small amounts of these elements in organic materials, and the bulk of that little has been concerned with aluminium and thallium.

Aluminium constitutes a considerable fraction of the earth's crust, yet

[1] Gutenmann, W. H., and Lisk, D. J., *J. Ag. Food Chem.* **8**, 306 (1960).
[2] Mulford, C. E., *Atomic Absorption Newsletter*, **5**, 135 (1966).

it occurs at only very low levels in plant and animal tissues. In plant materials the presence of any significant quantity of aluminium is generally taken to indicate soil contamination, and the determination of aluminium has been recommended as a measure of such contamination. The widespread use of vessels and equipment fabricated from aluminium alloys, inevitably leads to its presence in many foodstuffs, but its determination is rarely required.

Determinations of aluminium after destruction of the organic matter with both wet and dry oxidation procedures, have been reported, and the adverse comments noted have virtually all been applied to the dry ashing methods. Wet oxidations with any of the common digestion mixtures should be successful, and the choice will be determined mainly by the material to be decomposed. Dry ashing appears to have been used successfully at temperatures in the region of 500°C, although recovery data is sparse. In one survey[1] recoveries of 95 and 98% were reported for concentrations of 1 and 3 ppm after ashing carefully under conditions designed to prevent contamination of the sample: the ash was disolved in 6 N hydrochloric acid, and no difficulty was experienced, although higher ashing temperatures might be troublesome due to the increasing insolubility of the alumina produced. Other difficulties in recovering aluminium, have been reported,[2] when silica-containing samples were ashed. The ashes obtained were evaporated to dryness with hydrochloric acid, and then extracted with more acid. The dried residues were dissolved in hydrofluoric acid, and the retained aluminium determined. The amounts found by this procedure were very large, ranging to more than 2 mg per 5 g sample, and representing up to 85% of the aluminium present.

Contamination of samples with aluminium is said to occur when porcelain dishes are used for ashing, but a more serious problem is likely to arise from contamination of solutions handled or stored in glass apparatus or containers. Virtually all borosilicate glasses and bottle glasses contain appreciable amounts of alumina, ranging up to several per cent, and contamination from this source must be considered probable. For storage of solutions, a useful procedure is to line bottles with polythene bags, but this is clearly not possible for apparatus used at high temperatures.

[1] Thiers, R. E., in Glick, E., Ed., *Methods of Biochemical Analysis*, Vol. 5. Interscience, New York, 1957.
[2] Sandell, E. B., *Colorimetric Determination of Traces of Metals*. Interscience, New York, 1944.

1. Gallium and Indium

Very little has been published on the determination of these elements in biological materials, and very little recovery data is available.

One radiotracer study of the recovery of a number of elements, after oxidation with sulphuric acid and hydrogen peroxide included indium, and good recoveries (97 and 101 %) were reported. [1] As these conditions involved charring the sample with sulphuric acid the results might be taken as an indication that most of the common wet oxidation procedures are likely to be successful for indium. There is virtually no information available for gallium, but it will probably behave similarly to indium.

Compounds of both elements are fairly easily reduced, and it is possible that the occurrence of heavy charring during wet oxidation could lead to production of the metals. However, the boiling points of the elements are well over 1000°C in each case, so the vapour pressures, at the temperatures reached during digestion, are likely to be low.

A greater risk might exist during dry ashing, when higher temperatures are used, although ignition at 600°C has been reported with apparent success and comparison with other elements such as lead, whose behaviour is known, does not suggest that these temperatures are likely to cause serious trouble. The presence of chlorides which will yield hydrochloric acid at fairly high temperatures, might cause some difficulty, particularly with gallium because of the low temperature of sublimation of the trichloride. This problem might arise also with aluminium, although experience with iron under similar conditions has not shown it to be a serious cause of loss.

2. Thallium

Thallium is rather more important, in the context of this book, than gallium and indium, and more work has been reported on its determination in organic materials.

There would appear to be no reason to expect difficulty in the wet oxidation of samples for the recovery of thallium, and many such procedures have been reported without adverse comment.

Dry ashing has also been used, but with these techniques there is, perhaps, more risk. Thallium shows considerable similarities to lead and some of the problems experienced with the latter may well be found with thallium also. The main difficulties found in the recovery of lead were

[1] Down, J. L., and Gorsuch, T. T., *Analyst*, **92**, 398 (1967).

(i) losses by formation of silicates during dry ashing, (ii) losses by volatilization of the chloride from some types of sample during dry ashing, and (iii) coprecipitation of lead sulphate with calcium sulphate during wet oxidation of materials high in calcium, with mixtures containing sulphuric acid. The last of these is clearly not relevant in the case of thallium, as thallous sulphate is reasonably soluble, but the other two mechanisms might well apply. The oxide fairly readily forms a silicate, so loss by retention on silica might well occur, and the boiling point of thallous chloride is lower than that of lead chloride so that volatilization losses are also probable. However, this is largely surmise, and there is scope for further work on the subject. Unfortunately, the only convenient radioactive thallium tracer, thallium 204, is a beta emitter of moderate energy, so that the study of these problems would be rather more difficult than with a gamma emitting nuclide. However, this is a comparatively minor difficulty.

F. Scandium, Yttrium and the Rare Earths

There is little call for the determination of these elements in organic or biological materials, and the literature contains few references to suitable methods for separating them from such matrices. The main interest over the recent years has probably been in the determination of the radioactive isotopes produced by fission of uranium or plutonium and here the whole problem is greatly simplified by utilizing isotopic carrier techniques in which the low level of radioactive material is swamped with the inactive form of the same element, before the decomposition procedure is started. As the active material can readily be distinguished from the carrier, the final activity determination is unaffected, and provided that the whole system is fully equilibrated before there is any significant risk of loss, a comparison of the amount of inactive material recovered, with the amount known to have been added originally will give a correction factor equally applicable to the radioactive material.

Even apart from the simple situation found in radiochemical analysis, there seems to be no reason why these elements should give difficulty. Table 8.14 shows that either the melting points of the common salts are above the temperatures commonly used in ashing or the salts decompose to the very stable oxides. Similarly there should be no insolubility problems in wet digestion.

TABLE 8.14. MELTING POINT DATA

	Chloride	Nitrate	Sulphate	Oxide
Scandium	939	150	Decompose	
Yttrium	680		d1000	2410
Rare earths — range	590—890	Decompose	Decompose	

Dry ashing at temperatures up to 550 to 600°C should not cause trouble although the lowest efficient temperature is always the best. 600°C has been used with platinum dishes[1] but for silica or porcelain it is probably wiser to stay below this.

Wet oxidation with virtually any mixture of sulphuric, perchloric and nitric acids, or hydrogen peroxide should be perfectly satisfactory, with the efficiency of the mixture in dealing with the organic material being the main criterion for the choice.

The separation of yttrium from urine by acidification and absorption on a cation exchange resin has been described,[2] but this type of separation should only be used on samples where it has already been shown to work. All of these elements can form some very stable complexes with organic substances, and it cannot necessarily be assumed that a non-destructive treatment will be sufficient to release them.

G. The Actinide Elements
Actinium, Thorium, Protactinium, Uranium and Transuranics

The interest in this group of elements has increased markedly in the last twenty-five years, parallelling the numerous developments in the nuclear energy programme. The main interests have been in the radiological safety field, involving the determination of small quantities of these elements in biological tissues, generally as part of the various contamination monitoring programmes.

The initial stages of these determinations, involving the separation of the elements from the bulk of the organic materials, do not seem to have caused great difficulty, and there seems little point in referring to each of them separately: the findings with one element will probably apply to them all.

[1] Grant, C. L., *Anal. Chem.* **33**, 401 (1961).
[2] Schubert, J., Russell, E. R., and Farrabee, L. B., *Science*, **109**, 316 (1949).

Both wet and dry ashing methods have been used extensively for these determinations with little comment on their efficacy. In dry ashing temperatures at least up to 750°C have been used with apparent success[1] and the use of ashing aids such as nitric acid has been described. In wet ashing many combinations of the various commonly used oxidizing agents and strong acids have been employed, including the rarely used mixture of sulphuric acid and a catalyst.[2]

For urine samples, separation without destruction of organic matter has frequently been used, particularly for uranium and plutonium where coprecipitation with calcium phosphate in ammoniacal solution has been employed widely. In this type of analysis it has been shown[3] that although plutonium tracers, added to the urine could be obtained consistently at recoveries of greater than 90%, metabolized plutonium gave variable recoveries, some as low as 6%, unless the urine was first made acid with nitric acid and heated.

For normal samples there seems no reason to expect difficulty in recovering these elements by the normal wet or dry ashing procedures described in Chapter 9.

H. Germanium, Tin and Lead

All three of these elements are of tremendous technical importance, but with regard to their determination in organic materials there can be no doubt that the importance of tin and lead far outweighs that of germanium. This is clearly reflected in the number of investigations reported on each of the elements.

1. Germanium

A considerable volume of work has been described for the determination of germanium in coal. The method used for the preliminary decomposition of the sample has nearly always been dry ashing, generally without an ashing aid but sometimes with the addition of calcium oxide and calcium nitrate. The ashing temperatures used have often ranged up to 800°C.

[1] Petrow, H. G., and Strehlow, C. D., *Anal. Chem.* **39**, 265 (1967).
[2] Major, W. J., Wessman, R. A., Melgard, R., and Leventhal, L., *Health Phys.* **10**, 957 (1964).
[3] Shechan, W. E., Wood, W. R., and Kirby, H. W., U.S.A.E.C. Rep. TID 7696, 64 (1963).

An investigation into the factors affecting the loss of germanium from coal[1] has shown that the rate of heating is more important than the final temperatures attained. For example, at a final temperature of 550°C, household coal showed a loss of germanium of 10% when the heating rate was 20°C per minute, but only 2% when it was 3·5°C per minute. By contrast final temperatures between 400 and 1000°C all gave comparable results, provided that they were achieved at the same heating rate of 3·5°C per minute. Other points of interest arising from this were that loss of germanium appeared to be higher from coals of low density and ash content than from similar coals of higher density and ash content, and that the sulphur content of the coal had no effect on the extent of the losses.

Wet digestion methods have been used for the decomposition of coal, although some difficulties have been said to arise through the very firm absorption of germanium on silica during treatment with nitric and sulphuric acids. However, these same reagents have been considered the best for the preliminary oxidation of coal tar[2] where volatilization losses were found when the tar was dry ashed.

Similarly, in an extensive survey of the presence of germanium in the biosphere[3] normal dry ashing methods were discarded because of the risk of volatilization, and recourse was had to the low temperature ashing method of Gleit and Holland. Using this technique 5 mg additions of germanium oxide or sodium germanate were recovered at an average of 99·2%, and a spread of 86 to 105%.

By contrast, again, plant material has been ashed satisfactorily at 800°C with calcium oxide as an ashing aid.

Wet oxidation with sulphuric acid and hydrogen peroxide has been found to give reasonable recoveries when applied to organic materials not containing ionic or organically bound chlorine, but poor recoveries with those that do.[4] Some results from this radiochemical study are given in Table 8.15.

In view of this conflicting evidence it is difficult to draw firm conclusions as to the best methods to use for the preliminary destruction of carbonaceous materials before the determination of germanium.

[1] Menkovskii, M. A., and Aleksandrova, A. N., *Indust. Lab.* **29**, 853 (1963).
[2] Lechner, L., and Ferenczy, Z., *NehezvegyipKut Int. Kozlemen*, **2**, 353 (1963).
[3] Schroeder, H. A., and Balossa, J. J., *J. Chronic Diseases*, **20**, 211 (1967).
[4] Down, J. L., and Gorsuch, T. T., *Analyst*, **92**, 398 (1967).

TABLE 8.15. RECOVERY OF GERMANIUM AFTER SULPHURIC–PEROXIDE OXIDATION

Sample material	Recovery %	
None	97	92
Cocoa	69	48
Cocoa plus sodium chloride	10	13
Urea	100	101
Urea plus sodium chloride	12	9
Polythene	92	94
Polyvinyl chloride	3	3

The melting points and boiling points of some of the commoner inorganic compounds of germanium are set out in Table 8.16, and it can be seen that the forms likely to be volatile at normal ashing temperatures are the chlorides (and other halides), the monosulphide, and possibly the disulphide and monoxide.

TABLE 8.16. PHYSICAL DATA FOR SOME GERMANIUM COMPOUNDS

Compound		M.P.°C	B.P.°C
Sodium germanate	Na_2GeO_3	1083	
Germanium dichloride	$GeCl_2$	d. to $Ge + GeCl_4$	
Germanium tetrachloride	$GeCl_4$		84
Germanium dioxide	GeO_2	1086	
Germanium monoxide	GeO		Subl. 710
Germanium disulphide	GeS_2		Subl. 600
Germanium monosulphide	GeS		Subl. 430

In wet oxidation procedures there is some risk of loss of germanium whenever samples containing either ionic or covalent chlorine are heated in hot acid solution, and the results quoted in Table 8.15 do show very large losses. However, the tracer used in that case was germanium chloride in hydrochloric acid, so that it must be regarded as an extreme case. Further work is required, using tracers in other chemical forms, but losses under these conditions seem likely.

In dry ashing the situation is different. The presence of chlorine in the samples must again be regarded with suspicion, but in these circumstances ionic chlorine is likely to cause less trouble than covalent chlorine. The ready hydrolysis of the halides and the involatility of the germanates suggests that, in the presence of basic ashing aids, volatilization losses should not be serious.

2. Tin

Tin is a metal of great economic importance, particularly in the field of food preservation, and as such has attracted a significant amount of analytical attention. However, as the metal is non-toxic and as its level of occurrence in many foodstuffs is relatively high—in the tens of parts per million—the investigation has not been as searching as for, say, lead or arsenic.

The majority of methods described for the determination of tin in organic combination have involved wet oxidation procedures, mostly based on sulphuric acid. In a review of the chemistry of organo-tin compounds[1] the analytical section traces the development of the method through the use of nitric–sulphuric mixtures to digestion with sulphuric acid alone. The preferred method involves moistening the sample, contained in a Vycor crucible, with sulphuric acid, and heating the crucible with a ring burner, starting at the mouth, and working downwards towards the sample. This procedure is said to be suitable for volatile compounds although moistening of the sample with acetic acid is sometimes considered desirable.

Other wet oxidation procedures, using nitric and perchloric acids as well as sulphuric acid have been described and the use of excess nitric acid recommended, particularly with organo-tin compounds containing chlorine. For very volatile compounds a dry oxidation procedure has been developed[2] in which the sample is evaporated in oxygen through a heated plug of prepared asbestos where decomposition takes place.

For small amounts of tin in organic or biological materials, procedures involving either wet or dry oxidation have been widely used, although some doubts can be raised in each case.

The main complaint against dry ashing methods is that the tin is left in an insoluble form, probably SnO_2. To overcome this, fusion of the ash, often with a mixture of cyanide and carbonate, has been resorted to, although straightforward solution with hydrochloric acid has also, apparently been quite successful. The whole problem can, perhaps be

[1] Ingham, R. K., Rosenberg, S. D., and Gilman, H., *Chem. Rev.* **60**, 459 (1960).
[2] Brown, M. P., and Fowles, G. W. A., *Anal. Chem.* **30**, 1689 (1958).

avoided by merely charring the sample thoroughly, transferring the residue to a flask with sulphuric acid, and separating the tin by addition of hydrobromic acid to the hot solution, when tin is said to distil out. [1]

Wet digestion procedures have probably been the most widely used and although early workers believed that the use of nitric acid would lead to the precipitation of insoluble metastannic acid, it is, today, widely employed. Perchloric acid and hydrogen peroxide too, are commonly used in various combinations with sulphuric and nitric acids, but little recovery data is available. Radiochemical methods have shown that tin was not lost when cocoa, plus added sodium chloride, was first charred with sulphuric acid and then oxidized with hydrogen peroxide. [2] The tracer used in these experiments was in the stannous form, so that it is probable that all the chloride ion was lost before oxidation to the stannic form occurred. This system would be worth studying further, particularly with organic materials containing covalent chlorine, when chloride ion would be generated at a later stage in the digestion. Similarly, the recovery of tin, after oxidation with mixtures containing perchloric acid, would make an interesting study, as, in some ways, this would appear to be a singularly disadvantageous system. The presence of perchloric acid would ensure that the tin remained in the tetravalent state, while it has been shown[3] that with some types of organic material reduction of perchloric acid to chloride ion occurs at a relatively high temperature. This mechanism has been suggested as the cause of losses of mercury as mercuric chloride, so it seems even more likely to cause losses of tin as the much lower boiling stannic chloride. However, it must be stressed that this is conjecture, put forward in the hope that proof or rebuttal might be produced experimentally.

3. Lead

The bulk of the publications dealing with the determination of lead in organic materials have been concerned with public health problems, due to the toxicity, and cumulative effects, of the element and its compounds, and this concern has been given greater emphasis by the demonstration of the connection between ingestion of lead and mental retardation of children. This means that most of the work reported has been concerned with the

[1] Law, N. H., *Analyst*, **67**, 283 (1942).
[2] Down, J. L., and Gorsuch, T. T., *Analyst*, **92**, 398 (1967).
[3] Gorsuch, T. T., *Analyst*, **84**, 135 (1959).

determination of traces of lead, and although some organo-lead compounds are of great technical importance in the petroleum industry, only a fraction of the published papers deal with the initial degradation of organo-lead compounds prior to lead determination.

For non-volatile lead compounds many of the standard methods of decomposition have been recommended, such as wet oxidation with combinations of nitric and sulphuric acids and hydrogen peroxide on the one hand, and sulphated ashing, when the additional use of nitric acid is necessary to guard against the risk of reduction of the sulphate, on the other. Oxidation by the Carius method can also be used successfully with virtually all types of lead compound, but oxygen flask combustion, using the usual platinum sample support is not satisfactory as lead can be lost by formation of an alloy. [1]

In petroleum technology the additives in leaded petrols are usually treated with a halogen, generally bromine, as the first step in the determination. This is very satisfactory with pure compounds, and is the best method with volatile organo-lead compounds [2] but for highly unsaturated petrols the large consumption of bromine is a disadvantage. Under these circumstances, [3] agitation with concentrated nitric acid has been found adequate to convert the lead to the nitrate, and extract it into the aqueous phase, although wet oxidation of residual organic matter also extracted, may be necessary. More recently [4] the extraction has been carried out with hydrochloric, rather than nitric acid, although further oxidation of the residue is still required.

Some mixtures of lead tetra-alkyls have been separated gas chromatographically without any preliminary extraction procedure [5] and mixtures of di-, tri- and tetra-ethyl lead compounds and inorganic lead have been identified and determined by reaction with dithizone either directly, or, in the case of lead tetra-ethyl, after initial reaction with halogen. [6]

Turning to the far more numerous descriptions of methods used for preparing organic samples for the determination of traces of lead, we find both wet and dry oxidation methods extensively used, with the dry ashing methods, perhaps, slightly the more popular. However, it is notice-

[1] Belcher, R., MacDonald, A. M. G., and West, T. S., *Talanta*, **1**, 408 (1958).
[2] Leeper, R. W., Summers, L. S., and Gilman, H., *Chem. Rev.* **54**, 101 (1954).
[3] Baldeschweiler, E. L., *Ind. Eng. Chem. (Anal. Ed.)*, **4**, 101 (1932).
[4] Standard Methods of Test, ASTM 1964 *Book of Standards*, Part 17.
[5] Parker, W. W., Smith, G. Z., and Hudson, R. L., *Anal. Chem.* **33**, 1170 (1961).
[6] Henderson, S. R., and Snyder, L. J., *Anal. Chem.* **33**, 1172 (1961).

able that when difficulties are experienced, or losses found, it is nearly always the dry methods which are implicated.

Wet oxidation procedures, with nitric acid, sulphuric acid, perchloric acid and hydrogen peroxide used in most of the usual combinations, have all been considered satisfactory in the majority of applications. The results of some radiochemical recovery experiments for three of these combinations are given in Table 8.17.

TABLE 8.17. RECOVERY OF LEAD AFTER WET OXIDATION

Lead concn. ppm	Oxidation mixture	Organic material	Recovery % average	Ref.
1000	Nitric, sulphuric and perchloric acids	Vegetable matter	101	1
30–200	Nitric, sulphuric and perchloric acids	Animal matter	101	1
10	Nitric and perchloric acids	Cocoa	100	2
10	Nitric and sulphuric acids	Cocoa	92	2
10	Nitric, sulphuric, perchloric	Cocoa	96	2
10	Nitric and sulphuric acids	Dried milk	84	2

The most striking result is the very low recovery obtained when dried milk was digested with nitric and sulphuric acids. This was attributed to the coprecipitation of lead sulphate with the calcium sulphate formed from the large amount of calcium present in the milk powder, and the presence of lead in the precipitates was demonstrated by dissolving the precipitate and measuring the tracer released. The recoveries of lead from cocoa after wet oxidation with nitric and sulphuric acids, or nitric, sulphuric and perchloric acids are both rather low, but the recoveries of higher levels of lead from materials which probably contained rather less calcium than milk or cocoa appeared to be quite satisfactory. The results obtained with nitric and perchloric acids alone, where there is no risk of the formation of lead sulphate, were excellent. Generally speaking, mixtures containing sulphuric acid are likely to be satisfactory for samples with little or no calcium present, but may well cause trouble if the calcium contents are high.

[1] Pijck, J., Hoste, J., and Gillis, J., Int. Symp. Microchem., Birmingham, 1958.
[2] Gorsuch, T. T., Analyst, 84, 135 (1959).

Another problem reported to be associated with wet digestion procedures is the contamination of the samples with lead, during digestion in boro-silicate glass vessels.[1] This finding is supported[2] by other work which indicated contamination of purified water with 3 ppm of lead after storage in Pyrex vessels for two weeks. This type of problem is generally guarded against by pretreating glass equipment with the mixtures, and under the conditions, to be used in the determinations, but, in any event, lead oxide is not a quoted constituent of any of the normal borosilicate glasses, and Pyrex is stated to contain less than 5 ppm of lead.

Dry ashing methods for the destruction of organic material prior to the determination of lead have been reported with furnace temperatures ranging from 400°C or below to at least 800°C, although, as with most other elements, the commonest values are between 450 and 550°C.

The fact that losses of lead could occur during dry ashing procedures has been recognized for many years, and although the explanation most commonly advanced was that the losses were due to volatilization, it was recognized at least as early as 1930 that reaction could take place with the material of the ashing vessel, leading to retention losses. As with much of the work on recoveries of elements after ignition, the most comprehensive information is available from radioactive tracer studies, and some results are given in Table 8.18.

TABLE 8.18. RECOVERY OF LEAD AFTER DRY ASHING

Lead concn. ppm	Ashing temp.°C	Organic material	Ashing aid	Recovery % average	Ref.
60	400	Blood	None	103	3
60	500	Blood	None	69	3
60	700	Blood	None	32	3
60	900	Blood	None	13	3
1	450	Cocoa	None	98	4
1	550	Cocoa	None	95	4
1	650	Cocoa	None	77	4
1	450	Cocoa	$Mg(NO_3)_2$	98	4
1	550	Cocoa	$Mg(NO_3)_2$	93	4
1	650	Cocoa	$Mg(NO_3)_2$	94	4

[1] Hart, H. V., *Analyst,* **76,** 692 (1951).
[2] Healy, T. M., Morgan, J. F., and Parker, R. C., *J. Biol. Chem.* **198,** 305 (1952).
[3] Pijck, J., Hoste, J., and Gillis, J., Int. Symp. Microchem., Birmingham, 1958.
[4] Gorsuch, T. T., *Analyst,* **84,** 135 (1959).

Although the **agreement** between the two studies is not exact, clear trends can be seen. In the absence of ashing aid, the loss increases with increasing temperature, while the addition of magnesium nitrate reduces the high temperature losses. As the chemical form of the tracer can greatly affect the results of experiments of this kind, it should be noted that in ref. 4 (p. 96) the lead was added in the nitrate form. A more detailed examination of some results from ref. 4 is given in Table 8.19, and shows that the greater part of the losses indicated by the low recoveries can be accounted for as material attached to the silica crucibles used for the ignitions.

TABLE 8.19. DISTRIBUTION OF LEAD AFTER DRY ASHING
(In each experiment 1 ppm of lead was added to cocoa and ashed at 650°C.)

Ashing aid	Lead recovered %		Lead retained on crucible %	
None	71	83	28	15
H_2SO_4	96	90	2	2
$Mg(NO_3)_2$	91	96	4	2
$Mg(Ac)_2$	93	93	4	1

This indicates that with lead nitrate tracer, volatilization losses are not serious, but that reaction of the tracer with the crucible material can give a product very resistant to acid extraction. As it is suggested elsewhere in this book that the retention loss is due to the formation of a stable lead silicate it is clear that recoveries can be affected by the nature of the crucible, and the nature of the ashing aid, and some results covering these points are given in Tables 8.20, and 8.21.

As mentioned above, the nature of the tracer used in experiments of this kind can very greatly affect the results, and the use of lead chloride (M.P. 501°C) would be likely to demonstrate large volatilization losses.

TABLE 8.20. AMOUNTS OF LEAD RETAINED ON DIFFERENT CRUCIBLES

Nature of crucible	Lead retained % average	Number of measurements
Platinum	2	2
Old silica	16	8
New silica	28	2

TABLE 8.21. EFFECT OF ASHING AIDS ON BEHAVIOUR OF LEAD

Ashing aid	Temperature 650°C Lead recovered % average	Lead retained % average
None	69	23
H_2SO_4	71	29
H_3BO_3	101	0
H_3PO_4	40	43*
NaH_2PO_4	97	2
$Al(NO_3)_3$	99	0
$Mg(NO_3)_2$	100	0
NaCl	5	59*

* These figures are probably low.

This was shown to be the case when lead as chloride and as nitrate was added to organic matter and ashed at 650°C with magnesium nitrate as ashing aid. The recovery of the nitrate was 94%, but of the chloride only 46%. This increased volatility of lead chloride compared with lead nitrate raises the possibility that lead in any form might be lost by volatilization if it should be converted to the chloride when heated with inorganic chloride. This point has been explored[1] and it was shown that the presence of sodium chloride did not cause loss of lead by volatilization, but that ammonium chloride—which presumably dissociated to hydrochloric acid—did. Furthermore it was shown that the free energy changes associated with the reactions supported these findings.

$$PbO + 2NaCl = PbCl_2 + Na_2O \quad \triangle G_0 = + 73 \text{ kcal at } 600°C$$

$$PbO + 2HCl = PbCl_2 + H_2O \quad \triangle G_0 = -24 \text{ kcal at } 600°C$$

However, when lead tracer was heated with sodium chloride, only a very small part of the activity could be recovered by acid treatment (see Table 8.21) and it was shown that the remainder was adhering strongly to the crucible. It is therefore clear that the presence of sodium chloride in a sample being dry ashed for the recovery of lead, and indeed of most other elements, must be regarded with suspicion.

Another possible cause of variation in the recovery of lead is the nature of the organic material. It has been shown that materials containing

[1] Gorsuch, T. T., *Analyst*, **87**, 112 (1962).

covalent chlorine which give rise to hydrochloric acid during combustion—as, for example, polyvinyl chloride—can cause loss of lead by volatilization (see Table 8.22), whereas with other materials retention losses are more probable.

TABLE 8.22. THE EFFECT OF ORGANIC MATERIAL ON THE BEHAVIOUR OF LEAD TRACER
Ashing temperature 650°C; Lead nitrate tracer 1 ppm

Organic material	Lead recovered % average	Lead retained % average
Dried milk	83	15
PVC	1	1
Powdered tobacco	47	53

It is clear that dry ashing procedures for lead have a number of difficulties to overcome, hinging, in part, on the chemical form of the lead in the sample, and in part on the chemical nature of the sample itself. If any substantial quantity of chlorine is present it is probably easier and quicker to turn at once to the wet oxidation techniques, rather than embark on the investigation necessary to establish the suitability of a dry ashing procedure, although preliminary treatment with diluted sulphuric acid to drive off any ionic chloride may be a valuable alternative. For other samples ashing temperatures of about 500°C should be quite satisfactory, and the use of magnesium nitrate or aluminium nitrate as an ashing aid gives an additional safeguard.

One possible ashing aid which has not been widely used, boric acid, showed up remarkably well in tracer experiments, where, as well as allowing complete recovery of lead nitrate tracer, also allowed 100% recovery of lead chloride tracer, the only one of the ashing aids tested which did so. A possible explanation is advanced in the chapter covering the functions of ashing aids (Chapter 5).

I. Titanium, Zirconium and Hafnium

Of these three elements titanium is the most important commercially, being used as a constructional material and as a catalyst: zirconium also has some importance as a medium half-life (65 days) fission product.

Few problems have been reported in recovering these elements other than those arising from the refractory nature of the oxides which makes solution of the ash after dry ashing rather more difficult. This is generally overcome either by fusion of the ash with a suitable flux such as sodium carbonate, or potassium hydrogen sulphate, or by using a combined dry ashing and fusion process in which the fusion mixture is present from the beginning. Such a technique has been applied to the determination of titanium in polythene using a 3:1 mixture of potassium sulphate and potassium nitrate,[1] and to organic zirconium compounds using a tenfold excess of borax.

Wet oxidation methods have also been used, particularly in the analysis of plastics, where the usual procedure is to char the sample heavily with sulphuric acid, and then to clear it with a suitable oxidant such as nitric acid[1] or hydrogen peroxide.[2] A brief radiochemical investigation of the recovery of zirconium after a similar oxidation procedure, using sulphuric acid and hydrogen peroxide, showed complete recovery of 1 ppm of zirconium from cocoa containing added sodium chloride.[3]

J. Arsenic, Antimony and Bismuth

These three elements show a transition from non-metallic to metallic character with increasing atomic weight, and as they all form a range of organic compounds and as many of their compounds are poisonous, they are of considerable importance in this context.

The melting and boiling points of the elements, and of some of the most important compounds are given in Table 8.23.

1. Arsenic

Arsenic has the distinction of being, perhaps, the earliest element for which it was realized that destruction of organic matter was desirable prior to its determination: as early as 1839 Orfila was using potassium nitrate for this purpose.

The use of organo-arsenic compounds has been widespread, reaching into fields as different in intent as chemotherapy and chemical warfare, so

[1] Bolleter, W. T., *Anal. Chem.* **31**, 201 (1959).
[2] Novak, K., and Mika, V., *Chem. Průmysl*, **13**, 360 (1963) through *Anal. Abs.* **11**, 4414 (1964).
[3] Down, J. L., and Gorsuch, T. T., *Analyst*, **92**, 398 (1967).

TABLE 8.23. MELTING POINTS OF ELEMENTS AND COMPOUNDS °C

	Arsenic	Antimony	Bismuth
Element	sub. 615	630·5	271·3
Oxide X_2O_3	315*	656	820
Oxide X_2O_5	d. 315	d. 380	d. 150
Chloride XCl_3	− 8·5	73·4	231
Chloride XCl_5		2·8	

* Some forms sublime at lower temperatures.

for many reasons there has been extensive interest in the determination of arsenic in these compounds. The methods used have run the whole gamut, from refluxing with aqueous sodium bicarbonate to fusion with magnesium, with virtually every gradation of rigour occurring in between.

Dry methods have been used fairly frequently, generally with techniques involving closed systems. The catalytic reduction procedure described by ter Meulen in which the element is reduced to its hydride in a stream of hydrogen is clearly applicable to arsenic, and its use has occasionally been described, but more commonly oxidation methods have found favour. Simple combustion methods have been reported, but not very frequently, perhaps due to the uncertainties introduced by the varying volatility of the polymorphic forms of arsenious oxide.

Oxidation with oxygen under pressure, in a bomb, will not suffer from this disadvantage, in that it is relatively easy to wash out the whole volume of the bomb, but this method has not seen the widest application. More extensively investigated has been the oxygen flask method, the advantages of which appear, at first sight, to be quite marked. However, in the investigation of the application of the method to arsenic compounds, Corner[1] found that recoveries were low due to alloying of the arsenic with the usual platinum sample holder, and that it was necessary to use a silica spiral instead. Other workers[2] found that the use of paper impregnated with potassium nitrate reduced, but did not eliminate, this loss. As a sample holder, silica has never been found to be as satisfactory as platinum, as far as the combustion goes, and also appears to devitrify very readily.[3]

[1] Corner, M., *Analyst*, **84**, 41 (1959).
[2] Belcher, R., MacDonald, A. M. G., and West, T. S., *Talanta*, **1**, 408 (1958).
[3] Tuckerman, M. M., Hodecker, J. H., Southworth, B. C., and Fleischer, K. D., *Anal. Chim. Acta*, **21**, 463 (1959).

The final conclusion is, therefore that the oxygen flask method is not well suited to the determination of arsenic in organic compounds. [1] A substantial number of other "dry" methods have been described, including ashing with ashing aids such as magnesium nitrate, fusion with sodium peroxide, either alone or with other additions, and even fusion with magnesium to give the arsenide, but none of them has been widely applied. The greatest emphasis in this type of arsenic determination has been on the use of wet methods, and the usual wide range of mixtures can be found described in the literature. The advantages of the Carius method of sealed tube oxidation are clear, and it has been successfully applied, though not widely. Many other wet oxidation methods have been described, including some methods less commonly applied to other elements, as well as the usual combinations of sulphuric, nitric and perchloric acids. Ammonium persulphate has been used both in simple aqueous solution and in conjunction with either acid or alkali, and, even in simple solution, has been found satisfactory for materials as difficult as cacodylic acid. [2] More recently, a modification of this method has been applied to the analysis of rather exotic arsenic-containing polymers. [3]

A number of comparisons of the various wet oxidation methods have been made, but the results are not always consistent. Perhaps the most valuable is one based on experience of approximately seventy thousand arsenic determinations which considers eighteen different digestion procedures. [4] Unfortunately, the perchloric acid procedures were dismissed as dangerous, and not considered in detail, although many workers have used these procedures in safety. The ammonium persulphate procedures, using either water or sulphuric acid, were stated to be "highly" variable, and also not considered further.

Four types of digestion were extensively compared: sulphuric acid plus sodium sulphate, in the presence of carbohydrate; sulphuric acid and hydrogen peroxide; sulphuric and nitric acids plus sodium sulphate; and sulphuric acid plus potassium permanganate. The work was carried out in two parts: a study of the reproducibility of the methods for the determination of a single pure compound, and an investigation of the results obtained with a large range of compounds of different kinds. In the first

[1] Belcher, R., MacDonald, A. M. G., and West, T. S., *Talanta*, **1**, 408 (1958).
[2] Newbery, G., *J. Chem. Soc.* **127**, 1751 (1925).
[3] Hobin, T. P., *S.C.I. Monograph* No. 13, London, 1961.
[4] Banks, C. K., Sultzaberger, J. A., Maurina, F. A., and Hamilton, C. S., *J. Amer. Pharm. Assoc.* **37**, 13 (1948).

part of the study, the method using sulphuric acid and potassium permanganate was found to have an appreciably lower overall accuracy than the others, and was not examined further. The comparison of the results obtained with different arsenic compounds showed further differences between the remaining methods, with the sulphuric acid and sodium sulphate procedure failing in the presence of halogens, and both it and the peroxide and sulphuric acid method proving ineffective in liberating arsenic when it occurred as a heterocyclic ring element. For general use, excluding ring arsenic, it was considered that the peroxide and sulphuric digestion was the most satisfactory, provided that the peroxide was added before the acid; a technique which served to prevent interference by halogens. When arsenic was present in a ring compound only treatment with sulphuric acid and fuming nitric acid in the presence of sodium sulphate was found to be successful.

Even allowing for the extensive interest shown in the determination of arsenic in organic arsenical compounds themselves, it is probable that the major analytical concern has been with the determination of arsenic occurring in small amounts in other organic or biological materials. The importance of this element as a poison is reflected in the extensive legislation covering its occurrence in a wide range of materials, and its forensic significance is surely responsible for its popularity with writers of detective fiction. The minimum fatal dose in man is considered to be roughly 1 to 2 mg per kilogram of body weight, and as it has no known function in biological processes, its presence must be controlled to very low levels.

This need for the determination of small quantities of arsenic has led to a multiplicity of reports of methods used to prepare samples of organic matter for such determinations.

In some, perhaps rather specialized, instances, complete oxidation has been found unnecessary. For petroleum distillates, oxidative extraction with nitric and sulphuric acids under reflux was found to be sufficient to allow the remaining oil to be distilled away without seriously endangering the percentage recovery[1] while for rubber used for food containers a 24 hr extraction with 10% acetic acid at room temperature was deemed to be adequate.[2] As a more general procedure, acid hydrolysis of homogenized biological material using hydrochloric acid has been applied before

[1] Maranowski, N. C., Snyder, R. E., and Clark, R. O., *Anal. Chem.* **29**, 353 (1957).
[2] Moldrai, T., and Petrescu, G., *Industria uscara*, **12**, 522 (1965).

liberation of arsenic as arsine. Reduction with stannous chloride was then found to proceed quantitatively. [1]

Despite the simplicity and apparent convenience of such relatively non-destructive methods, the majority of workers have found it desirable, or even essential, to turn to the more rigorous, and more widely used, total destruction procedures.

As might be expected, virtually all the usual oxidation procedures have been applied to the recovery of arsenic, with a pronounced preference for wet oxidation methods being apparent. Many comparisons of various methods have been carried out, and although, almost inevitably, some conflict of evidence is to be found, some conclusions can eventually be drawn.

Turning first to consideration of methods used for dry ashing, a number of recovery experiments have been recorded, and quite violent conflict between the results is to be seen. Most of the recoveries reported have been after dry ashing the material without the use of ashing aids, although the use of numerous aids is also described in the literature. Results of some of the recovery experiments are given in Table 8.24.

The results taken from refs. 1–3 (p. 105) were obtained using chemical recovery methods, while the remainder (refs. 4–6) employed radiochemical methods. With the exception of the results marked with an asterisk(*), the experimental procedures involved the addition of the arsenic tracer to the sample shortly before ashing, with the attendant limitations discussed earlier. In the marked results, the organic matrices themselves were labelled, by injecting rats with the radioactive tracer, so that the arsenic had time to become incorporated within the tissues before the rats were sacrificed. It is clear that the results are far from consistent, but some interesting trends can be observed. The nature of the organic material appears to have a direct bearing on the amount of arsenic lost, a finding which, in itself, might explain some of the inconsistencies noted. Of the methods studied, the use of magnesium nitrate as an ashing aid gives the most favourable indications, with several investigators finding recoveries near to 100%.

One type of dry oxidation which needs to be considered separately, is the low temperature method using excited oxygen. [2] The original publication indicated very good recoveries from blood spiked with inorganic

[1] Kingsley, G. R., and Schaffert, R. R., *Anal. Chem.* **23**, 914 (1951).
[2] Gleit, C. E., and Holland, W. D., *Anal. Chem.* **34**, 1454 (1962).

TABLE 8.24. RECOVERY OF ARSENIC AFTER DRY ASHING

Arsenic level	Temperature °C	Ashing aid	Organic matrix	Recovery %	Reference
		$Mg(NO_3)_2$		100	1
1 ppm		$Mg(NO_3)_2$	Liver	78	2
1 ppm	600	$Mg(NO_3)_2$	Potato	68–116	3
22 ppm	400	None	Blood	23	4
22 ppm	500	None	Blood	0	4
22 ppm	700	None	Blood	0	4
	450	None	Blood	72	5
	550	None	Blood	71	5
	850	None	Blood	57	5
	450	None	Bone	51	5*
	450	None	Blood	14	5*
	450	None	Kidney	18	5*
1 ppm	550	None	Cocoa	88	6
1 ppm	550	HNO_3	Cocoa	84	6
1 ppm	550	$Mg(NO_3)_2$	Cocoa	99	6
1 ppm	550	H_2SO_4	Cocoa	96	6

arsenic tracer, but a later study,[7] using atomic absorption spectrophotometry, found that arsenic losses varied with the power supplied to the r.f. coil, ranging from nought at minimum power to 90% at maximum power. The sample used was highly artificial, but it is clear that further work is required, preferably with samples containing arsenic which is fully incorporated into the sample itself.

Despite the great volume of work carried out in investigating the recovery of arsenic during dry ashing, it is apparent that much remains to be done. Studies such as that of Hamilton, Minski and Cleary, in which the radioactive tracer is fully incorporated into the sample need to be extended, with both animals and plants being used, and with the effect of the very many possible ashing aids being investigated.

The wet ashing procedures have, on the whole, given more consistent results, although there is still far from unanimous agreement.

[1] Cassil, C. C., *J.A.O.A.C.* **20**, 171 (1937).
[2] Allcroft, R., and Green, H. H., *Biochem. J.* **29**, 824 (1935).
[3] Lisk, D. J., *J. Ag. Food Chem.* **8**, 121 (1960).
[4] Pijck, J., Hoste, J., and Gillis, J., Int. Symp. Microchem., Birmingham, 1958.
[5] Hamilton, E. L., Minski, M. J., and Cleary, J. J., *Analyst*, **92**, 257 (1967).
[6] Gorsuch, T. T., *Analyst*, **84**, 135 (1959).
[7] Mulford, C. E., *Atomic Absorption Newsletter*, **5**, 135 (1966).

The recovery of arsenic after oxidation of organic matter by the usual combinations of nitric, sulphuric and perchloric acids, and hydrogen peroxide have been investigated radiochemically, and the results are given in Table 8.25.

TABLE 8.25. RECOVERY OF ARSENIC AFTER WET OXIDATION

Arsenic level	Oxidation mixture*	Organic matrix	Mean recovery%	Reference
1 ppm	N + P	Cocoa	99	1
1 ppm	N + P + S	Cocoa	99	1
1 ppm	N + S	Cocoa	98	1
1 ppm	S + H$_2$O$_2$	Polythene	98	2
1 ppm	S + H$_2$O$_2$	Cocoa + NaCl	60	2
1 ppm	S + H$_2$O$_2$	PVC	3	2

* N + P = nitric and perchloric acids.
N + P + S = nitric, perchloric and sulphuric acids.
N + S = nitric and sulphuric acids
S + H$_2$O$_2$ = sulphuric acid and 50% hydrogen peroxide.

Generally speaking, the mixtures containing perchloric acid are accepted as giving good recoveries of arsenic, although some workers reject them on other grounds.

Mixtures containing only sulphuric and nitric acids, on the other hand, are less completely accepted, and doubts have been raised concerning their efficacy, at least in certain rather specialized circumstances. Gross[3] reported that even fuming acids would not destroy all the heterocyclic compounds present in tobacco leaves, and that some of these interfered with the determination of arsenic so that the apparent recoveries were low, even though the arsenic was present in solution. Other workers have stated that it is necessary to maintain oxidizing conditions at all times, particularly if chloride is present, in order to prevent volatilization of arsenic as the trichloride. This last point has been tested both chemically, and radiochemically, and would appear, under the normally used conditions, to be relatively unimportant. The vital point seems to be that the sample must be heated in the presence of excess nitric acid first, and only allowed to

[1] Gorsuch, T. T., *Analyst*, **84**, 135 (1959).
[2] Down, J. L., and Gorsuch, T. T., *Analyst*, **92**, 398 (1967).
[3] Gross, C. R., *Ind. Eng. Chem. (Anal. Ed.)*, **5**, 58 (1933).

char when the chloride ion has been removed as nitrosyl chloride. If the material is charred with sulphuric acid alone, the presence of chloride ion is likely to cause very serious losses of arsenic (see Table 8.25).

The use of sulphuric acid and hydrogen peroxide for the decomposition of samples containing arsenic is relatively recent, and the results quoted in Table 8.25 show that the nature of the sample can have a very substantial effect. The almost complete loss of arsenic demonstrated when polyvinyl chloride was oxidized, is fairly clearly due to the formation of arsenic trichloride during the stages when heavy charring with sulphuric acid occurred, and similar losses are likely to be found during the oxidation of any compound containing covalent chlorine. Conversely, the negligible loss when polythene was decomposed indicates that organic materials not containing chlorine can be successfully decomposed without loss of arsenic. Falling between these two types of material are samples containing substantial amounts of chloride ion, but no covalent chlorine. The study quoted above, showed that large losses do occur when samples of this kind are charred initially with sulphuric acid, before treatment with hydrogen peroxide, but it seems probable that dilution of the sulphuric acid, and one or two preliminary evaporations prior to charring would remove the bulk of the chloride ion as hydrochloric acid, and reduce or even prevent such volatilization losses.

It is desirable that further studies into the recovery of arsenic after wet oxidation should be made using samples containing "natural" radioactive arsenic, that is samples containing radioactive arsenic tracer incorporated into them, rather than merely added prior to the experiment. The ready availability of arsenic 74, with a half-life of 18 days, makes this approach much more feasible than it was when the only accessible arsenic tracer was arsenic 76 with its short half-life of $26 \cdot 5$ hr.

Of the generally applied methods for the decomposition of organic materials prior to the determination of arsenic, the most effective are probably the ones employing mixtures containing perchloric acid. Using these under the recommended conditions, where the material is never allowed to char, loss of arsenic is likely to be minimized from all but the most refractory samples. For people unwilling to use perchloric acid mixtures, nitric and sulphuric acids would seem to be the best choice. The major difficulty here is with materials containing chlorine, particularly when it is present in covalent form. Chloride ion can generally be removed in the manner discussed above, but covalent chlorine poses the substantial difficulty that it cannot be removed until it has been converted to chloride ion, and in some instances the procedure necessary to ensure its conversion to chloride (e.g. charring with sulphuric acid) will itself lead to losses by volatilization. This problem is at its most severe with the chlorine-containing plastics, which are resistant to lesser forms of attack such as nitric acid or hydrogen peroxide, and only succumb to the action of hot concentrated sulphuric acid. With these materials even the perchloric acid mixtures are less than satisfactory unless the plastic is first partially degraded, when, of course, the same problems arise as with nitric and sulphuric acids.

The use of hydrogen peroxide and sulphuric acid is also likely to be very effective with non-chlorine containing samples, as evidenced by the results relating to the destruction of polythene. With samples containing chloride, preliminary treatment with water or hydrogen peroxide in the presence of sulphuric acid, but without preliminary charring, will probably be satisfactory, but this procedure needs to be tested on specific materials: for samples containing covalent chlorine all the arguments listed above apply, and for materials as resistant to attack as PVC, a solution is hard to find. The most promising procedure, with any of the oxidizing mixtures, would be to use a closed system in which the distillate is trapped for further treatment.

Hydrogen peroxide, particularly the more concentrated forms, also has a useful supporting role to perform in completing the destruction of any residual material remaining after decomposition with nitric and sulphuric acids. It is possible that such a treatment with 50% hydrogen peroxide would overcome the difficulties experienced by Gross, but this would require verification.

The suitability of dry oxidation methods for the recovery of arsenic is, at present, somewhat uncertain. Ashing with magnesium nitrate appears to be the most satisfactory of the many procedures described, but more rigorous testing is required, using a wide variety of different samples.

In discussing the recovery of arsenic from biological material the work of Satterlee and Blodgett is of considerable significance. [1] They found that blood and tissue contained a thermolabile, arsenic-containing fraction which was lost on drying at temperatures as low as 56 to 80°C. If such substances are of common occurrence the use of dry methods of oxidation for these materials are immediately eliminated, and even wet oxidation

[1] Satterlee, H. S., and Blodgett, G., *Ind. Eng. Chem.* (*Anal. Ed.*), **16**, 400 (1944).

procedures would require further examination. In this field the use of "naturally" labelled samples would seem to be highly desirable for further progress.

2. Antimony

Although organo-antimony compounds do not have the same significance as the organo-arsenicals, there has still been a reasonable amount of attention paid to the determination of antimony in them. The usual types of wet oxidation method have been used with apparent success, particularly those in which treatment with sulphuric acid is combined with the action of an oxidizing agent such as potassium permanganate, hydrogen peroxide or nitric acid. Fusion with sodium carbonate, soda lime and sodium peroxide have also been described, although the last method, using a Parr bomb has been said to convert the antimony to an obdurately insoluble form, which made further treatment very difficult. [1]

The determination of traces of antimony in large samples of organic material has generally been required because of its importance as a poison. The oxidation procedures used have been many and various, and some difficulties have been experienced.

Dry ashing procedures, at temperatures up to 550°C, have generally been considered satisfactory, and the use of magnesium nitrate has been widely recommended. This view has been supported by one set of radiotracer recovery experiments in which recoveries, of 1 ppm of antimony from cocoa, ranged between 92 and 97%, but has been contradicted by other tracer results which showed a recovery of only 67% of the antimony when blood spiked with 11 ppm of antimony was ashed at a temperature as low as 400°C. The results of these two studies are given for comparison in Table 8.26.

Such large differences may possibly be explained by differences in the nature of the organic matrix used, but are more reasonably due to the nature of the tracer used. If the tracer used in the experiments with blood consisted of antimony chlorides in strong hydrochloric acid, then some loss by volatilization, might well be expected. In the experiments with cocoa the tracer was prepared by dissolving antimony trioxide in sodium hydroxide to give a much less volatile form. The differences between the two sets of tracer results emphasize once again the need for caution in interpreting the figures obtained with this technique, and underline the

[1] Wilkinson, N. T., *Analyst*, **78**, 165 (1953).

TABLE 8.26. THE RECOVERY OF ANTIMONY AFTER DRY ASHING

Organic material	Ashing aid	Antimony concentration ppm	Ashing temperature °C	Recovery %	Reference
Cocoa	None	1	550	96	1
Cocoa	HNO₃	1	550	92	1
Cocoa	H₂SO₄	1	550	94	1
Cocoa	Mg(NO₃)₂	1	550	97	1
Blood	None	11	400	67	2
Blood	None	11	500	82	2
Blood	None	11	700	35	2
Blood	None	11	900	9	2

fact that, unless special steps are taken to equilibrate the tracer with the element already present in the sample, the results obtained can often be varied, almost at will, by varying the chemical nature of the tracer used.

It has sometimes been suggested that antimony might be lost when samples are ashed which contain substantial amounts of chloride ion, particularly if present as sodium chloride. This has been checked radio-chemically, using sodium chloride and ammonium chloride,[3,4] and it has been found that although the presence of ammonium chloride led to very large losses of antimony by volatilization (approximately 90%) the presence of sodium chloride had no such effect. Moreover, it was demonstrated that the thermodynamics of the reactions were fully in accord with these findings. However, it was also shown that in the presence of sodium chloride very large amounts of antimony, in the region of 50%, were firmly retained by the silica crucibles used for the ashing, so that, although volatilization was not the cause, very substantial losses were brought about by the action of sodium chloride.

Turning to wet oxidation procedures, the results of the two tracer investigations discussed above are more nearly in accord, with, on the one hand, recoveries varying from 94 to 100% after the digestion of cocoa with nitric and perchloric, nitric and sulphuric, and nitric, perchloric and sulphuric acids, and, on the other, recoveries varying from 94 to 100% after

[1] Gorsuch, T. T., *Analyst*, **84**, 135 (1959).
[2] Pijck, J., Hoste, J., and Gillis, J., Int. Symp. Microchem., Birmingham, 1958.
[3] Gorsuch, T. T., *Analyst*, **87**, 112 (1962).
[4] Gorsuch, T. T., Ph. D. Thesis, London University, 1960.

the oxidation of a variety of materials with nitric, perchloric and sulphuric acids. However, despite these encouraging figures problems have been found which were due, not to the physical loss of antimony, but to its conversion to a form in which it was not amenable to the detection method used. This led some workers to the conclusion that losses were occurring, but Maren[1] showed that, during oxidation of organic material with nitric and sulphuric acids, some of the antimony was converted to a form, said to be four-valent, in which it would neither react with rhodamine B, nor oxidize further to the five-valent state in which it would react. He also found that the use of perchloric acid in the final stages of the oxidation ensured the complete conversion of the antimony to the reactive form.

The other major wet oxidation procedure, using sulphuric acid and hydrogen peroxide, has also been studied radiochemically[2] and good recoveries obtained, even in the presence of large amounts of sodium chloride. However, the final valency state of the antimony was not demonstrated in these experiments, so that it cannot be stated dogmatically that problems similar to those explored by Maren would not arise. Further work would be required to answer these questions unequivocally.

From the published work it would appear that the best general method for the destruction of organic matter for antimony determinations would be one involving a final perchloric oxidation. Sulphuric acid and hydrogen peroxide, or sulphuric acid and nitric acid would probably serve equally well if the final determination were to be made by a method other than the colorimetric procedure using rhodamine B. Despite the contradictory results reported in Table 8.26, it is also probable that dry ashing with magnesium nitrate as an ashing aid, will give satisfactory results.

3. Bismuth

This element has attracted substantially less attention than either arsenic or antimony, and little has been reported in the way of difficulty in the destruction of organic matter.

In a review of organo-bismuth chemistry[3] the use of sulphuric and nitric acids was considered the most satisfactory digestion procedure, but all the usual combinations of nitric, sulphuric and perchloric acid, and of hydrogen peroxide have been used with apparent success.

[1] Maren, T. H., *Anal. Chem.* **19,** 487 (1947).
[2] Down, J. L., and Gorsuch, T. T., *Analyst,* **92,** 398 (1967).
[3] Gilman, H., and Yale, H. L., *Chem. Rev.* **30,** 281 (1942).

A similar tale can be found in the literature on the determination of traces of bismuth, where all the common wet oxidation techniques appear to give satisfactory results, and where dry ashing at temperatures of about 500°C, with or without ashing aids, also seems to be acceptable.

One tracer study, into recoveries of a variety of elements after digestion with sulphuric acid and hydrogen peroxide, included bismuth, and showed no losses after the oxidation of cocoa. [1]

K. Vanadium, Niobium and Tantalum

Although niobium and tantalum have aroused relatively little general interest in this field, the determination of vanadium in organic materials is of considerable significance. This is mainly due to its importance as a poison, both toxicologically and industrially, and it is in the petroleum industry, where vanadium in the feed to cracking plants has an adverse effect on the yields of petrol, that the majority of determinations are carried out. The amounts of vanadium vary widely, but in some cases nearly a quarter of the ash may be in the form of V_2O_5.

The importance of the determination of vanadium in petroleum products, has led to the development of rapid methods of analysis, of which X-ray fluorescence has been the most widely used although electron spin resonance and neutron activation analysis have also been considered.

The techniques used for the destruction of the oil prior to vanadium determinations have generally been variants of a dry ashing procedure, with or without a variety of additives. The information available on the likelihood of loss of the element during the procedures is somewhat contradictory, and although special techniques have been described to prevent loss, these have not always been considered necessary. Comparative figures for the recovery of vanadium after the application of three different ashing procedures to a Venezuelan gas oil indicated little difference between the methods used. [2] The procedures were:

(i) Simple burning followed by ashing at 550°C.
(ii) Simple burning, followed by treatment with sulphuric acid and ashing.
(iii) Treatment with sulphuric acid followed by burning and ashing.

[1] Down, J. L., and Gorsuch, T. T., *Analyst*, **92**, 398 (1967).
[2] Barney, J. E., and Haight, G. P., *Anal. Chem.* **27**, 1285 (1955).

Similar results were obtained by other workers who found that for residues and crude oil there was no difference in recovery after simple ashing or sulphated ashing, [1] and they further stated that vanadium compounds in petroleum are of high molecular weight and low volatility. This general thesis of the non-volatility of vanadium compounds gains added support from experiments with synthetic vanadium tetraphenyl porphyrin [2] where again simple and sulphated ashing procedures gave comparable results.

A number of more specialized procedures have been described, designed either to speed up the rate of sample preparation, or to reduce the possibility of losses. One of the most interesting is the use of benzene sulphonic acid to destroy the vanadium porphyrin complexes. [3] This allows ignition at a temperature of 600 to 700°C which is well below the temperature at which volatilization of V_2O_5 will occur, [4] although there still exists the possibility of reaction with the material of the ashing vessel. Another recommended procedure involves the addition of 10% of sulphur to the oil sample to reduce losses. [5]

The use of oxygen under pressure in a calorimetric bomb to remove all possibility of volatilization loss has been described [6] but the example has not been widely followed. Neither has there been much application of a very different technique in which the oil sample is dropped on to a heated silica–alumina catalyst at a slow rate, so that it is cracked and vaporized almost immediately, and the catalyst mixture finally heated at 600°C [7] Although this was found satisfactory for the spectrographic procedure to which it was applied, it is unlikely to be successful for ordinary chemical determinations, as the conditions seem almost ideal for causing large losses by solid state reaction between the catalyst and the constituents of the ash.

Wet oxidation procedures are not generally popular for petroleum products due to the small samples that can be successfully handled, and

[1] Anderson, J. W., and Hughes, H. K., *Anal. Chem.* 23, 1359 (1951).

[2] Horeczy, J. T., Hill, B. N., Walters, A. E., Schutze, H. G., and Bonner, W. H., *Anal. Chem.* 27, 1899 (1955).

[3] Shott, J. E., Garland, T. J., and Clark, R. O., *Anal. Chem.* 33, 506 (1961).

[4] Milner, O. I., Glass, J. R., Kirchner, J. P., and Yurick, A. N., *Anal. Chem.* 24, 1728 (1952).

[5] Agazzi, E. J., Burtner, D. C., Crittenden, D. J., and Patterson, D. R., *Anal. Chem.* 35, 332 (1963).

[6] Rayner, J. E., *J. Inst. Fuel*, 37, 30 (1964).

[7] McEvoy, J. E., Milliken, T. H., and Juliard, A. L., *Anal. Chem.* 27, 1869 (1955).

although the use of several of the common combinations of sulphuric, nitric and perchloric acids and hydrogen peroxide have been reported, there does not seem to be a strong case for their use prior to the determination of vanadium.

The literature on the determination of vanadium in materials other than petroleum products is very much more scanty, but few difficulties seem to have arisen. Graphite has been ashed at temperatures up to 650°C after first treating it with sodium hydroxide[1] and various wet oxidation procedures have been applied successfully. Indeed there is no reason to doubt that any of the common combinations could be used on appropriate samples. A radiochemical investigation of the recovery of vanadium from organic samples, with and without the addition of sodium chloride, after oxidation with sulphuric acid and hydrogen peroxide, showed recoveries of 98 to 101%,[2] while decomposition of blood, urine, and other animal and vegetable tissue with nitric, perchloric and sulphuric acids repeatedly showed 100% recoveries.[3]

The other two elements in this section, niobium and tantalum are of little significance in the context of this monograph. Niobium has some importance due to the occurrence of niobium 95 as a fission product of medium half-life (35 days) but generally interest has been small. Neither element is likely to be lost by volatilization during the oxidation of organic material and the greatest problems are likely to be solution of the oxides after ignition, and the need to maintain adequate concentrations of complexing agents to keep them in solution.

L. Chromium, Molybdenum and Tungsten

The destruction of organic material prior to the determination of these three elements is, on the whole, a straightforward matter, with relatively few difficulties being reported: determination of tungsten has been required only infrequently.

Wet oxidation seems to be quite satisfactory, and most of the common mixtures should give good results. Radio tracer investigations with chromium at 10 ppm and molybdenum at 1 ppm have shown excellent recoveries after oxidation of organic matter with nitric and sulphuric

[1] Kul'berg, L. M., and Shilova, A. G., *Uch. Zap. Saratov. Univ.* **42,** 71 (1955) through *Anal. Abs.* **4,** 1182 (1957).

[2] Down, J. L., and Gorsuch, T. T., *Analyst*, **92,** 398 (1967).

[3] Pijck, J., Hoste, J., and Gillis, J., Int. Symp. on Microchem., Birmingham, 1958.

acids, nitric, sulphuric and perchloric acids and nitric and perchloric acids. [1]

A similar study has demonstrated the complete recovery of 1 ppm of chromium, whether added as chromic ion or as chromate, after oxidation with sulphuric acid and hydrogen peroxide. [2]

One conceivable source of loss during oxidation of organic materials containing chromium, might be due to the formation of volatile chromyl chloride (B.P. 117°C). This requires the presence of chloride ion, and although any chloride originally present in the sample will probably distill off during the first minutes of the oxidation, there are at least two mechanisms by which chloride ion can be generated at high temperature: organic materials containing covalent chlorine, such as PVC, may decompose, or, in oxidations involving perchloric acid, materials very resistant to oxidation may reduce this acid at elevated temperatures. Despite these plausible mechanisms, such losses of chromium do not appear to have been reported. Indeed, when an attempt was made to induce such volatilization by oxidizing ethanolamine, which is known to be resistant to oxidation, with nitric, sulphuric and perchloric acids, no chromium was distilled even though copious production of chloride ion was clearly demonstrated. [3] This was the case whether the chromium tracer was added as chromic chloride or as sodium chromate. A possible explanation of these findings may be found in the fact that chromyl chloride is readily decomposed by water, and the water content of the oxidizing medium may have been too high at the time that the chloride ion was generated.

Dry ashing has also been widely used as the preliminary oxidation stage for these elements, although a number of difficulties have been reported. Radiochemical experiments have shown good recoveries of both chromium (10 ppm) and molybdenum (1 ppm) after ashing at 550°C in silica. [1] Other tracer studies [4] show similar recoveries for larger amounts of chromium and molybdenum (66 to 140 μg) at temperatures up to 500°C, but a fall in the recovery at higher temperatures, down to 56 and 83% respectively at 900°C. Other workers have used dry ashing with various ashing aids, and apparent success, at temperatures in the general region of 500 to 600°C.

[1] Gorsuch, T. T., *Analyst*, **84**, 135 (1959).
[2] Down, J. L., and Gorsuch, T. T., *Analyst*, **92**, 398 (1967).
[3] Gorsuch, T. T., Ph. D. Thesis, London, 1960.
[4] Pijck, J., Hoste, J., and Gillis, J., Int. Symp. Microchem., Birmingham, 1958.

However some difficulties have been encountered, even at low ashing temperatures, with recoveries of molybdenum of only 70% after 12 hr ashing at 450°C[1] and even lower recoveries, down to 24% after ashing at 500°C.[2] It has also been reported that the use of nitric acid as an ashing aid can cause loss, by volatilization, of MoO_3, at temperatures below 450°C but the experience of other workers makes this seem unlikely.

A tracer study of the reaction of chromium on heating to ashing temperatures (600°C) in silica baisins, with sodium chloride or ammonium chloride has shown that although none of the element is lost by volatilization in the presence of sodium chloride, up to half of the chromium can be lost by retention on the silica.[3] Although these conditions were highly artificial, with no organic matter present, they do indicate one possible mechanism to explain the losses reported above.

M. Manganese, Technetium and Rhenium

Manganese is much the most important of these three elements in the context of this monograph, and virtually the only one for which any discussion is possible.

Technetium does not occur naturally, being known mainly because of its occurrence in fission products, while rhenium, though naturally occurring, is very rare. Both elements have volatile heptoxides; technetium is known to distil from sulphuric acid solutions containing nitric or perchloric acids[4] and it is probable that rhenium will do so also, so that the usual wet oxidation procedures are not advisable. Perhaps the best approach would be ashing with an alkaline ashing aid, although it is possible that direct dry ashing would be successful, due to the reducing action of the charred organic matter.

Manganese, by contrast, has attracted an appreciable amount of attention, and both wet and dry oxidation methods have been applied with relatively little adverse comment. Dry ashing temperatures up to at least 800°C have been used, but tracer recovery experiments have shown losses of 15% at 700°C and 20% at 800°C,[5] so that upper limits of 500 to

[1] Smit, J., and Smit, J. A., *Anal. Chim. Acta*, **8**, 274 (1953).

[2] Scharren, K., and Munk, H., *Agrochimica*, **1**, 44 (1956).

[3] Gorsuch, T. T., Ph.D. Thesis, London, 1960.

[4] Ishibashi, Masayoshi, Fujinaga, Taitiro, and Koyama, Mutsuo, *J. Chem. Soc. Japan Pure Chem. Sect.* **81**, 1260 (1960).

[5] Pijck, J., Hoste, J., and Gillis, J., Int. Symp. on Microchem., Birmingham, 1958.

550°C seem to be advisable. Loss of manganese by reaction with silica in the sample is sometimes considered to be a hazard, and the use of a low maximum temperature will help to minimize this type of loss.

Wet oxidations with mixtures of nitric, perchloric and sulphuric acids, and hydrogen peroxide have all been applied successfully, and there seems to be no reason to expect difficulty.

A collaborative intercomparison, among seventeen laboratories, of wet oxidation and dry ashing at 550°C, prior to the determination of manganese in feeds, revealed no significant differences between the two techniques, for samples containing from 50 to nearly 400 ppm of manganese. [1]

N. Selenium, Tellurium and Polonium

These elements form a sub-group of increasing metallic character, and varying analytical significance. Selenium is important toxicologically, while polonium is important radiologically, and the differences between the three elements make it convenient to treat them separately.

1. Selenium

Selenium closely resembles sulphur in many ways, and forms an extensive range of organic compounds. Some of these are chemically hard to distinguish from their sulphur analogues, and follow very similar metabolic pathways: for example selenomethionine is readily incorporated into proteins as a substitute for methionine.

Many methods have been used for the separation of selenium from organic materials, and considerable difficulties have been experienced at all scales of working.

For the analysis of organo-selenium compounds, the use of routine combustion train techniques is said to be unsatisfactory, as the selenium formed tends to foul the packings. [2] A satisfactory method appears to be combustion in oxygen, followed by passage over platinum to form selenium dioxide which is absorbed in water. [3]

Perhaps an obvious technique to apply to the determination of an element as troublesome as selenium, is the Carius oxidation in a sealed

[1] Heckman, M., *J.A.O.A.C.* **50**, 45 (1967).
[2] Campbell, T. W., Walker, H. G., and Coppinger, G. M., *Chem. Rev.* **50**, 279 (1952).
[3] Hallett, L. T., *Ind. Eng. Chem.* (*Anal. Ed.*), **14**, 956 (1942).

tube, and this has often been applied successfully. An early survey of suitable methods for the determination of selenium in organic compounds[1] considered this to be the most satisfactory of all in the absence of halogens, while later workers have used it even for compounds containing halogens.[2]

Similarly, other techniques involving completely closed systems have been used, including fusion with sodium peroxide in a microbomb, and burning in oxygen in a closed flask, both of which appear to give satisfactory recoveries.

A rather more vigorous development of the normal combustion procedure has been described, in which the sample is heated in a furnace in a stream of hydrogen, and the hydrogen, carrying the decomposition products, subsequently ignited in oxygen.[3] This method is said to be suitable for the decomposition of —O—Se— linkages but this claim has been disputed[4] on the grounds that a residue remained which was hard to oxidize without near explosive violence. Instead a wet oxidation procedure in which the sample was heavily charred with sulphuric acid and then further oxidized with fuming nitric acid was recommended. Similar systems have been described by other workers, but in view of the serious losses found in trace analysis, which are discussed later, this type of oxidation would seem to be suspect.

Oxidizing mixtures containing perchloric acid have not been widely used for the destruction of organo-selenium compounds, although they would seem, from work involving the determination of traces of selenium, to be well suited to this application.

For trace analysis, wet oxidation procedures have found much wider acceptance than dry ashing methods, due to the readiness with which selenium is lost during ignition.

The use of dry ashing in the presence of magnesium nitrate after a preliminary treatment with nitric acid, has been described[5] and it is possible that the combination of initial oxidizing treatment and the use of magnesium nitrate is adequate to prevent reducing conditions developing. In general, however, dry ashing methods have been found to cause

[1] Bradt, W. E., and Lyons, R. E., *Proc. Indiana Acad. Sci.* **36**, 195 (1926).
[2] Cook, H. G., Illett, J. D., Saunders, B. C., and Stacey, G. J., *J. C. S.* **1950**, 3125.
[3] McCullough, J. D., Campbell, G. W., and Krilanovitch, N. J., *Ind. Eng. Chem. (Anal. Ed.),* **18**, 638 (1946).
[4] Gould, E. S., *Anal. Chem.* **23**, 1502 (1951).
[5] Oelschläger, W., *Landw. Forsch.* **18**, 79 (1965) through *Anal. Abs.* **13**, 6950 (1966).

serious losses unless they are carried out in closed systems. Radiochemical experiments with a combined wet and dry ashing procedure, in which the maximum temperature attained was only about 350°C, still showed that 1 ppm of selenium tracer was completely lost during the ashing procedure. [1]

One form of open dry oxidation system for which good recoveries of selenium have been reported is the low temperature excited oxygen procedure, [2] where 99% recovery of selenium after the oxidation of alfalfa, was reported. This favourable result was not, however, confirmed by other workers who found that selenium could not be quantitatively recovered from pure cellulose, when present alone, but could in the presence of some other elements, provided that the power input levels were kept low. [3] Possibly the poor recoveries were related to the virtually ashless organic matrix employed.

The closed dry ashing system most commonly used for samples of this kind has been the oxygen flask technique, using flasks of up to about 10 litres capacity. It has been said that the residue sometimes remaining after the samples have burned away can contain small amounts of selenium, [4] and to improve the completeness of the combustion and to prevent the formation of such a residue a method has been described in which animal samples are mixed with magnesium nitrate and pelleted before ignition. [5]

Most oxygen flask procedures start with a dried ground sample which is then burnt, but a recent paper [6] indicates that normal drying methods can lead to the loss of volatile selenium compounds. To overcome this problem, these workers freeze dried the samples in cellophane containers before combustion and, under these circumstances, quoted recoveries of from 85 to 104%.

Because of the problems of normal dry ashing, and the sample size limitations of the oxygen flash technique, most of the reported work on the determination of traces of selenium has been carried out after wet oxidation of the samples. The standard A.O.A.C. procedure, which has been widely used throughout the world, involves partial oxidation with

[1] Gorsuch, T. T., *Analyst*, **84**, 135 (1959).
[2] Gleit, C. E., and Holland, W. D., *Anal. Chem.* **34**, 1454 (1962).
[3] Mulford, C. E., *Atomic Absorption Newsletter*, **5**, 135 (1966).
[4] Watkinson, J. H., *Anal. Chem.* **38**, 92 (1966).
[5] Lane, J. C., *Irish J. Agric. Res.* **5**, 177 (1966).
[6] Taussky, H. H., Washington, A., Zubillaga, E., and Milhorat, A. T., *Microchem. J.* **10**, 470 (1966).

nitric and sulphuric acids, in the presence of mercuric oxide as a fixative, and distillation of selenium as the bromide. This method has been extensively tested [1] with experiments carried out in which $0 \cdot 1$ to $6 \cdot 7$ ppm of selenium were added to a variety of samples, and recoveries of $95 \cdot 5$ to $102 \cdot 6\%$ obtained.

Despite these good results, considerable conflict is evident regarding the use of sulphuric and nitric acids alone. Some workers recommend methods in which the sample is heated with sulphuric acid or sulphuric and nitric acids until charring occurs, while others say that losses are to be expected unless oxidizing conditions are maintained throughout. Radiochemical investigations [2] have shown that losses of selenium during oxidations with nitric and sulphuric acids can be related to the extent to which the organic material is allowed to char; slow oxidation with no charring gave recoveries of 95 and 98 %, while rapid oxidation with heavy charring gave recoveries of only 3 and 4 %. It was assumed that the onset of charring indicated the occurrence of reducing conditions.

The clear lesson of this finding is that it is necessary to maintain oxidizing conditions at all stages of wet oxidations, and this requirement explains the efficiency of oxidizing mixtures containing perchloric acid. Radio tracer studies [2,3] have shown that both nitric and perchloric acids, and nitric, sulphuric and perchloric acids gave good recoveries of selenium when used for the oxidation of organic matter. This is presumably related to the high boiling point of perchloric acid which ensures that it is not driven off in the early stages of the digestion, so that oxidizing conditions are maintained at all times. This contrasts with the situation when nitric acid is used when all of this acid can be volatilized long before the oxidation of the organic matter is complete. Many oxidations have been described using perchloric acid mixtures, prior to the determination of selenium, and it is probable that such mixtures are the most satisfactory for this work. It is only with perchloric acid that the twin goals of complete oxidation of the sample and quantitative recovery of the selenium can be achieved, although it is worth noting that even with a nitric–perchloric mixture serious losses have been reported when the sample was taken to dryness. [4]

One paper has investigated recoveries of selenium after oxidation of

[1] Klein, A. K., *J.A.O.A.C.* **24**, 363 (1941).
[2] Gorsuch, T. T., *Analyst*, **84**, 135 (1959).
[3] Kelleher, W. J., and Johnson, M. J., *Anal. Chem.* **33**, 1429 (1961).
[4] Stanton, R. E., and McDonald, A. J., *Analyst*, **90**, 497 (1965).

plastic materials with sulphuric acid and hydrogen peroxide, and recoveries of 66 and 72% from PVC and 11 and 16% from polythene, were obtained. [1] This presumably reflects the much greater amount of carbon and hydrogen in the latter leading to the maintenance of reducing conditions over a longer period of time.

2. Tellurium and Polonium

Few papers are to be found on the determination of these elements in organic materials, but some general comments are possible. Dry ashing has been shown to be quite unsatisfactory for polonium, as for selenium, and there is every reason to suppose that the same will apply to tellurium. By the same token it is to be expected that mixtures of perchloric acid with nitric or nitric and sulphuric acids will be perfectly successful for the recovery of both these elements.

O. Iron, Cobalt and Nickel

These three elements are all of great technical significance and their occurrence in organic materials is widespread and important. However the roles they play differ significantly, and investigations into methods of separation have tended to deal with one element or another, rather than with the three of them together, so that it is in many ways simpler to consider them separately here.

1. Iron

Iron is widely spread in biological materials, and is important in them in a number of respects. It is a relatively large component of the human body, having an overall concentration of some 40 to 50 ppm, and a level of about 500 ppm in blood. It is an essential trace element for many plants, and is found in very many foods, partly naturally, and partly adventitiously, entering from the iron and steel used for construction, packing, etc. Its presence in food can have undesirable effects, such as its action in catalysing oxidative rancidity in fats, or in the production of a haze in some liquids, and it is also an objectionable impurity in other technical products, such as petroleum.

As might be expected with such an important element, a considerable amount of work has been carried out on its separation and determination.

[1] Down, J. L., and Gorsuch, T. T., *Analyst*, **92**, 398 (1967).

With some samples in which the iron is not strongly complexed, the metal can be determined virtually without any preliminary treatment. Simple salts such as the acetate, can be titrated directly with a chelating agent such as EDTA, or dissolved in glacial acetic acid and titrated as a base with perchloric acid.[1] Liquid samples, either wholly organic such as acrylonitrile[2] and light coloured glyceride oils,[3] or aqueous solutions such as wine,[4] can sometimes be treated directly with chromogenic reagents, and the colours so developed measured either in the sample itself or after extraction into a suitable solvent.

Instrumental methods such as atomic absorption spectrophotometry and X-ray fluorescence spectrometry are finding increasing use for the direct determination of iron in a variety of materials. Atomic absorption is particularly useful for petroleum products which can often be dissolved in a suitable solvent and sprayed into the flame, although this technique falls down if the iron is contained in particulate material of uneven size. This objection does not apply to X-ray fluorescence which is reasonably sensitive for iron, and has been applied to a range of organic materials including oils,[5] plastics[6] and plant materials[7] at levels down to 0·1 ppm.

At a more rigorous level, iron can be extracted from many organic materials by an acid extraction. This has been applied very widely to blood samples (where treatment with hydrochloric acid is often coupled with the use of a protein precipitant such as trichloracetic acid) and to petroleum products (where some quite complex extraction procedures have been described)[8] but its application is not limited to these materials.

When it has been necessary to achieve complete destruction of the organic matter prior to the determination of iron, both wet and dry methods have been widely used. For organo-iron compounds, where there is no other mineral matter present, the standard textbooks are virtually unanimous in recommending ashing in a stream of air or oxygen to give ferric oxide,

[1] Casey, A. T., and Starke, K., *Anal. Chem.* **31**, 1060 (1959).
[2] Maute, R. L., Owens, M. L., and Slate, J. L., *Anal. Chem.* **27**, 1614 (1955).
[3] Newlove, T. M., in Cocks, L. V., and Van Rede, C., Eds., *Laboratory Handbook for Oil and Fat Analysts*, p. 346. A. P. London, 1966.
[4] Collins, P., and Diehl, H., *Anal. Chim. Acta*, **22**, 126 (1960).
[5] Dwiggins, C. W., and Dunning, H. N., *Anal. Chem.* **32**, 1137 (1960).
[6] Westrik, R., Int. Cong. on the Appl. of X-ray Spec. to Ind. 8/10 June 1960 (Liege).
[7] Jenkins, R., and Hurley, P., *Analyst*, **91**, 395 (1966).
[8] Barney, J. E., *Anal. Chem.* **27**, 1283 (1955).

and weighing the product, although the use of nitric acid, nitric acid and peroxide, sulphuric acid and peroxide, and nitric and sulphuric acids have all been reported for various organic compounds of iron.

For the determination of small amounts of iron in various organic or biological materials, there has been widespread use of both wet oxidation and dry ashing. Wet oxidation has proved very successful in almost all cases, with most of the possible combinations of nitric, sulphuric and perchloric acids, and hydrogen peroxide being used. Recovery experiments by various workers, have nearly always given good and consistent results. Radiochemical measurements of iron at the 1 ppm and 10 ppm level, using combinations of nitric, sulphuric and perchloric acids have shown recoveries between 97 and 102% [1] in good accord with other radiochemical work which showed recoveries of 96 to 106% in the 30 to 400 ppm range [2] and ordinary chemical determinations which gave overall recoveries of $100 \mp 1\%$. [3] Mechanical losses, eliminated by the use of a spray trap, have been reported [4] but most criticisms of the wet oxidation methods have been minor, such as the tedious nature of the operation itself, or the need to remove reagents such as hydrogen peroxide or perchloric acid to prevent interference with subsequent determinations. A serious exception to this general agreement on the suitability of wet digestion methods is to be found in a collaborative study by members of the A.O.A.C. in which six feed samples were analysed for iron (among other elements) after destroying the organic matter by the A.O.A.C. method 22·070 [5] using nitric and perchloric acids. The iron contents were in the range 100 to 1200 ppm and the results obtained by the seventeen collaborators were considered to be of unsatisfactory precision, with coefficients of variation ranging up to 28%. [6] By contrast, the coefficients of variation when ash solutions were distributed to the participants so as to eliminate sample preparation errors, were approximately 8%. Not surprisingly, further work on this topic was suggested.

Despite this disturbing finding, the weight of evidence is that wet oxidation methods are satisfactory for the recovery of iron (see Table

[1] Gorsuch, T. T., Ph. D. Thesis, Univ. of London, 1960.
[2] Nicholas, D. J. D., Lloyd-Jones, C. P., and Fisher, D. J., *Plant and Soil*, **8**, 367 (1957).
[3] Jackson, S. H., *Ind. Eng. Chem. (Anal. Ed.)*, **10**, 302 (1938).
[4] Aubouin, G., *Radiochimica Acta*, **1**, 117 (1963).
[5] *Official Methods of Analysis*, 10th Edn. A.O.A.C. 1965.
[6] Heckman, M., *J.A.O.A.C.* **50**, 45 (1967).

8.27). Indeed it is difficult to postulate mechanisms, other than mechanical entrainment, by which the element could actually be lost.

However, it has recently been demonstrated that an oxidation method employing sulphuric, nitric and perchloric acids, which involved fuming the final oxidized solution, led to the formation of anhydrous ferric sulphate which was insoluble in anhydrous sulphuric acid. When samples containing more than 50 mg of iron were oxidized and the iron determined immediately after dilution, recoveries were always low, falling to less than 20% when 200 mg of iron were present. Dilution of the acid mixture and boiling for ten minutes restored the situation completely. This point must obviously be borne in mind, but the usual procedure of diluting the digest and boiling to remove oxides of nitrogen should deal with this problem also.

The methods described in Chapter 9 using various combinations of nitric, sulphuric and perchloric acids, and hydrogen peroxide should be suitable for a wide variety of samples, although for materials high in fat it has been claimed that preliminary treatment with hydrogen peroxide and sulphuric acid, followed by nitric acid gives more rapid digestion. [1]

TABLE 8.27. RECOVERY OF IRON AFTER WET OXIDATION

Sample	Iron content	Method	Recovery %	Reference
Cocoa	1–10 ppm	Radiochemical	97–102	2
Plant material	30–400 ppm	Radiochemical	96–106	3
Synthetic	10 ppm	Chemical	100 ∓ 1	4
Beer	0·2–1 ppm	Chemical	99–101	5
Tomato juice	1–5 ppm	Chemical	99–100	5
Coffee	0·2–1 ppm	Chemical	95–101	5

Turning now to the use of dry ashing methods prior to the determination of iron, we find that much more controversy is evident. Many workers have used such methods with apparent satisfaction, or at least without

[1] Thompson, J. B., Ind. Eng. Chem. (Anal. Ed), 16, 646 (1944).

[2] Gorsuch, T. T., Ph. D. Thesis, Univ. of London, 1960.

[3] Nicholas, D. J. D., Lloyd-Jones, C. P., and Fisher, D. J., Plant and Soil, 8, 367 (1957).

[4] Jackson, S. H., Ind. Eng. Chem. (Anal. Ed.), 10, 302 (1938).

[5] Roberts, H. L., Beardsley, C. L., and Taylor, L. V., Ind. Eng. Chem. (Anal. Ed.), 12, 365 (1940).

comment, while some have reported quite adequate recoveries. On the other hand many variations in recovery with the nature of the sample, the nature of the ashing aid and the temperatures used are to be found in the literature.

Satisfactory recoveries have been reported by a number of workers after ashing samples directly, without the use of an ashing aid, generally at temperatures ranging from 450 to 550°C. Some results are summarized in Table 8.28.

TABLE 8.28. SATISFACTORY RECOVERIES OF IRON AFTER DRY ASHING

Temperature °C	Iron level ppm	Recovery %	Method of determination	Reference
450–500	Approx. 100	97–103·5	Chemical	1
500 ∓ 50	20–1000	94–106	Chemical	2
350–700		97·8	Radiochemical	3
250–450	10 + 30	92	Spectrographic	4
550	10–350	Approx.100	Radiochemical	5
550	10	99–101	Radiochemical	6

In addition to these results it is worth noting that in the collaborative work of the A.O.A.C. mentioned above, in which wet oxidation was found to be unsatisfactory, the coefficients of variation for five of the six feed samples, after dry ashing were between 4·8 and 9·5%.

However, to set against these favourable reports, there are many instances of difficulties occurring during or after the dry oxidation of samples.

One radiochemical investigation[7] found a recovery of only 86% after ashing blood containing 140 μg of iron at 400°C for 24 hr and only 82, 52 and 27% after shorter periods at 500, 700 and 900°C. This work emphasizes the need to keep the temperature low, although the poor

[1] Hummel, F. C., and Willard, H. H., *Ind. Eng. Chem. (Anal. Ed.)*, **10**, 13 (1938).
[2] Wyatt, P. F., *Analyst*, **78**, 656 (1953).
[3] Joyet, G., *Nucleonics*, **9**, (6) 42 (1951).
[4] Thiers, R. E., in Glick D. Ed., *Methods of Biochemical Analysis*, Vol. 5, p. 273. Interscience, New York, 1957.
[5] Morgan, L. O., and Turner, S. E., *Anal. Chem.* **23**, 1365 (1951).
[6] Gorsuch, T. T., *Analyst*, **84**, 135 (1959).
[7] Pijck, J., Hoste, J., and Gillis, J., Int. Symp. on Microchem., Birmingham, 1958.

recovery at 400°C is surprising. An apparently even more critical dependence on temperature is to be found in a report on the determination of iron in foods[1] in which maximum recovery was obtained after ashing for 5 hr at 550°C. A longer or shorter time, or a higher or lower temperature were all said to reduce the recovery.

A valuable solution to the problem of losses due to high temperatures is provided by the use of electrically excited oxygen for low temperature decomposition.[2] Using this technique a recovery of 101% was obtained, and the maximum temperature reached was less than 100°C.

Many comparisons of the effects of various ashing aids on the recovery of iron have been carried out, and considerable variation reported.

The use of sodium hydroxide was found to increase the recovery of iron from as little as 47 to 100% when ashing breadmaking materials at "a low redness".[3] As an incidental in this work it can be seen that the recovery of iron varied from one material to another after ashing under constant conditions (see Table 8.29).

TABLE 8.29. RECOVERY OF IRON FROM DIFFERENT MATERIALS

Organic material	Recovery %
"Special cereal"	25
Molasses	75
Whole wheat bread	80

Another investigation, concerned with the recovery of iron from biological material,[4] found all the ashing aids tried to be unsatisfactory. The use of sulphuric acid gave recoveries up to 57%, calcium carbonate was slightly better at 80 to 87%, but sodium carbonate was so poor that the use of it was abandoned at an early stage.

A considerable body of comparative work is available from the petroleum industry on the efficiency of various ashing procedures. The standard ashing method involves igniting the sample and allowing it to burn to a carbonaceous residue before ashing it in a furnace; the most

[1] A. Iregui Borda, *Rev. Colombianaquim,* **3,** 53 (1949).
[2] Gleit, C. E., and Holland, W. D., *Anal. Chem.* **34,** 1454 (1962).
[3] Hoffman, C., Schweitzer, T. R., and Dalby, G., *Ind. Eng. Chem. (Anal. Ed.),* **12,** 454 (1940).
[4] Jackson, S. H., *Ind. Eng. Chem. (Anal. Ed.),* **10,** 302 (1938).

common alternative is to add a quantity of sulphuric acid to the sample before burning to act as an ashing aid. Little difference was found between the two techniques when oxidizing crude oils and residues[1] but considerably improved recoveries were obtained by sulphated ashing when distillates were analysed. This was considered reasonable because only volatile iron compounds would distil into the overhead fractions in the first place. These findings were confirmed by later work[2] in which iron tetraphenyl porphyrin was synthesized and added to oil before ashing. Again the sulphated ashing gave better recoveries than the standard method. However, a further comparison using distillate fractions[3] found no difference between simple ashing and sulphated ashing both of which gave good recoveries, but, rather surprisingly found that if sulphuric acid is added to the sample after burning, but before ignition in the furnace, the recovery of iron fell to about sixty per cent.

The low values often reported for the recovery of iron after the dry ashing of organic materials have been attributed to a number of causes. All the mechanisms proposed fall into one of two groups, either loss by volatilization or loss by reaction with some solid material.

Volatilization losses can be divided in turn, into two further categories, losses due to volatilization of some volatile compound already present in the sample, as has been suggested for the iron porphyrins in petroleum distillates, or loss of a volatile species produced during the ashing process. The most common example quoted in this latter category is the formation of ferric chloride by reaction of the iron compound with sodium chloride and warnings against this are contained in many standard texts. However, work with radioactive tracers has shown that not only does this mechanism not operate, but that it is not thermodynamically feasible.[4] The free-energy change, at 600°C, for the reaction

$$Fe_2O_3 + 6\ NaCl = 2\ FeCl_3 + 3Na_2O$$

is $+$ 313 kcal, so that there is no likelihood of losses arising from this cause.

On the other hand, the presence of sodium chloride can greatly influence the losses due to the reaction of iron with solid material present in the

[1] Milner, O. I., Glass, J. R., Kirchner, J. P., and Yuric, A.N., *Anal. Chem.* **24,** 1728 (1952).

[2] Horeczy, J. T., Hill, B. N., Walters, A. E., Schutze, H. G., and Bonner, W. H., *Anal. Chem.* **27,** 1899 (1955).

[3] Barney, J. E., and Haight, G. P., *Anal. Chem.* **27,** 1285 (1955).

[4] Gorsuch, T. T., *Analyst,* **87,** 112 (1962).

system. This is particularly the case where the ashing is carried out in silica or porcelain dishes when the presence of chloride ion can greatly weaken the silicate structure and increase the reaction of the iron with it. This has been demonstrated[1] in a series of experiments in which iron 59 tracer was heated in silica crucibles with sodium chloride, when up to 40% of the iron was retained by the crucibles. This type of loss has also been described by Petersen[2] who considered that it became important at temperatures above 600°C. He also found that porcelain was more resistant to this type of attack than were silica or Vycor, although other workers have decided against porcelain, either because of low recoveries[3] or contamination.[4]

Even without the addition of sodium chloride the reaction of iron with silica has often been quoted as a cause of loss. Sandell[5] quotes a paper by Wijkstrom in which hay and straw samples containing various amounts of silica were ashed, the residue carefully extracted with hydrochloric acid and the iron still remaining on the residue measured. After ashing in a muffle furnace, large amounts of iron, totalling hundreds of micrograms were found to be retained on the residues from 5 g samples, with the greatest retention losses occurring in the samples with the highest silica contents.

As with most retention losses encountered in dry ashing procedures the best solution is to keep the ashing temperature as low as possible, and to introduce an inert diluent material into the final ash by adding an ashing aid, such as magnesium nitrate. If, despite these measures, serious losses by reaction with silica still occur, the only solution is to transfer the ash to a platinum baisin and volatilize the silica by heating with hydrofluoric acid.

Another serious cause of error in the determination of iron after dry ashing is due to the formation, in the ash, of complex polyphosphates which can sequester the iron and render it unavailable to the reagent used to determine it.[2,4] The most usual solution to this problem is to hydrolyse the phosphates by dissolving the ash in hydrochloric acid and evaporating the solution slowly to dryness on a water bath, before redissolving the

[1] Gorsuch, T. T., Ph. D. Thesis, London, 1960.

[2] Petersen, R. E., *Anal. Chem.* **24**, 1850 (1952).

[3] Joyet, G., *Nucleonics*, **9**, (6) 42 (1951).

[4] O'Connor, R. T., Heinzelman, D. C., and Jefferson, M. E., *J. Am. Oil. Chem. Soc.* **24**, 185 (1947).

[5] Sandell, E. B., *Colorimetric Determination of Traces of Metals*. Interscience, New York, 1944.

residue. This treatment has not always been considered satisfactory for materials high in phosphorus[1] and further treatment of the residue obtained by evaporation of the acid solution of the ash, with ammonium sulphide solution, was found to be necessary.

An observation made in the field of mineral analysis, which has some relevence to this discussion, is the finding by Shell[2] that if a sodium carbonate fusion of rocks is carried out over a gas flame, the conditions are sufficiently reducing to cause considerable loss of iron by reduction to the metal and alloying with the platinum of the crucible. In a muffle furnace with free circulation of air, this loss does not occur. As the early stages of a dry ashing, even in a muffle furnace, must give rise to reducing conditions this mechanism must be considered as a possible cause of error in subsequent determinations. However, it has been shown by Ellingham[3] that the lowest temperature at which ferrous oxide is reduced to the metal by carbon is about 700°C, so that at the usual temperatures used for dry ashing, reduction losses are not a serious hazard. However, this is yet another argument for not using high temperatures.

2. Cobalt

Cobalt is an element of considerable economic importance, particularly as an essential trace element whose absence has caused serious disease among sheep in several parts of the world. It is, therefore, not surprising that most of the publications on the subject deal with its determination in animal tissues or other biological materials, and the majority of the methods described for the destruction of organic material prior to cobalt determination refer to the destruction of relatively large amounts of sample.

When organic compounds containing large percentages of cobalt are to be handled the standard textbooks are in complete accord in recommending gravimetric determination after reduction to the metal, either directly in a current of hydrogen, or after combustion in air or oxygen. One or two mention determination of cobalt as the oxide. Decomposition of organo-cobalt compounds with concentrated hydrochloric acid has been described[4] while simple compounds can often be determined directly

[1] Woiwood, A. J., *Biochem. J.* **41**, 39 (1947).
[2] Shell, H. R., *Anal. Chem.* **26**, 591 (1954).
[3] Ellingham, H. J. T., *J. Soc. Chem. Ind.* **63**, 125 (1944).
[4] Ingles, D. L., and Polya, J. B., *J. Chem. Soc. Lond.* 1949, 2280.

by titration, either acidimetrically in non-aqueous solvents [1,2] or by complex formation, [3] or even, in specialized cases, absorptiometrically. [4]

The determination of small quantities of cobalt in biological materials has not excited a great deal of comment on the methods used to separate it from the organic matter. Some liquid samples can be handled without any separation procedures, as with the colorimetric determination of cobalt in beer [5] while others, such as some oil products, need just a simple extraction. [6] X-ray fluorescence procedures have been applied to the determination of cobalt in organic matrices, and under suitable conditions have obvious advantages. In general, however, and particularly in view of the very low levels of cobalt often involved, most determinations do necessitate the preliminary oxidation of the sample, and both wet and dry methods have been widely used. Adverse comments have been fairly limited and a considerable variety of oxidation procedures has been recommended.

Dry ashing has the pronounced advantage, for all low-level determinations, that the reagent blank can be kept to a minimum, and for this reason it is a desirable procedure for cobalt. Excellent recoveries have been reported [7] down to the millimicrogram level, using ashing at 450°C in a

TABLE 8.30. RECOVERY OF COBALT FROM ORGANIC MATERIAL

Organic material	Cobalt level	Ashing aid	Temperature °C	Recovery	Reference
Blood	$m\mu g$	None	450	Complete	7
Cocoa	1 ppm	None	550	99%	8
Cocoa	1 ppm	HNO_3	550	99%	8
Cocoa	1 ppm	H_2SO_4	550	99%	8
Cocoa	1 ppm	$Mg(NO_3)_2$	550	101%	8
Blood	~40 ppm	None	400	98%	9

[1] Casey, A. T., and Starke, K., *Anal. Chem.* **31**, 1060 (1959).

[2] Rashbrook, R. B., *Analyst*, **87**, 826 (1962).

[3] Graske, A., *Off. Dig. Fed. Soc. Paint Technol.* **33**, 855 (1961).

[4] Malik, W. U., and Haque, R., *Z. Anal. Chem.* **189**, 179 (1962).

[5] Stone, I., *Proc. Am. Soc. Brew. Chem. 1965*, p. 151.

[6] Herhausen, E. P. R., and DeGray, R. J., *Ind. Eng. Chem. (Anal. Ed.)*, **14**, 806 (1942).

[7] Thiers, R. E., Williams, J. F., and Yoe, J. H., *Anal. Chem.* **27**, 1725 (1955).

[8] Gorsuch, T. T., *Analyst*, **84**, 139 (1959).

[9] Pijck, J., Hoste, J., and Gillis, J., Int. Symp. on Microchem., Birmingham, 1958.

filtered air stream and these favourable results have been supported by radiochemical experiments at higher levels (see Table 8.30).

Even higher temperatures have also apparently been used with success, for example Saltzmann[1] ashed biological materials at up to 700°C, but some workers have not found the method satisfactory. Radiochemical investigations at temperatures of 500, 700 and 900°C revealed recoveries of only about 70% in each case when ashing blood containing about 40 ppm of cobalt,[2] and normal inactive recovery experiments at a temperature of only 450°C have shown losses of about 30%.[3] Radio tracer studies, in which 20 μg amounts of labelled cobalt, as the nitrate, were heated in silica crucibles at 630°C, with or without the addition of organic material, showed that large percentages of cobalt can be retained on the silica when there is no organic matter present, but that the addition of even a virtually ashless organic material reduced the retention to an average of 3%.[4] A similar effect has been even more strikingly demonstrated in radiochemical experiments using porcelain crucibles, where recoveries ranged only from 1·6 to 10·9% when 10 mg amounts of cobalt salts were heated alone for 2 hr at 600°C,[5] but when similar amounts of radioactive cobalt were added to a feedstuff and ashed, recoveries of 96 to 97% were obtained. Here again the presence of diluting material, preventing close contact between the cobalt and the crucible, was cited as the reason for the improvement. The addition of a bulky ashing aid such as magnesium nitrate would therefore be expected to reduce such retention even further, but the evidence in the literature, for the efficacy of various ashing acids is a little confused. For liver, the use of calcium acetate, calcium hydroxide, magnesium acetate and magnesium nitrate were all found to be unsatisfactory[6] while for plant tissue the use of simple ashing was found to be better than ashing with nitric acid.[7] On the other hand, for the oxidation of pasture materials, ashing after treatment with nitric or sulphuric acids was found to give the best results.[8]

[1] Saltzmann, B. E., *Anal. Chem.* **27**, 284 (1955).
[2] Pijck, J., Hoste, J., and Gillis, J., Int. Symp. on Microchem., Birmingham, 1958.
[3] Smit, J., and Smit, J. A., *Anal. Chim. Acta,* **8**, 274 (1953).
[4] Gorsuch, T. T., *Analyst,* **84**, 139 (1959).
[5] Mays, D. L., Kessler, W. V., Chilko, J. A., and Schall, E. D., *J.A.O.A.C.* **50**, 735 (1967).
[6] Kidson, E. B., Askew, H.O., and Dixon, J. K., *N.Z. J. Sci. Tech.* **18**, 601 (1936/7).
[7] Hiscox, D. J., *Sci. Agric.* **27**, 136 (1947).
[8] McNaught, K. J., *Analyst,* **67**, 97 (1942).

Despite the conflicting evidence it is probable that dry ashing methods are quite satisfactory for low levels of cobalt, provided that low temperatures are used. There is evidence that at temperatures of 600°C or above some cobalt is lost by interaction with silica and probably porcelain, basins, although this should not be the case with platinum. Temperatures of 450°C should be satisfactory, and the use of a little nitric acid, to speed the oxidation, or magnesium nitrate to supplement low ash contents, would probably be advantageous in many cases.

Various wet ashing methods have been described on many occasions with little comment on their efficiency either relative to each other or to the numerous dry ashing variants. There is no obvious reason why serious losses should occur with any of the commonly used mixtures, and the criteria governing the choice of a method are likely to be convenience and its ability to destroy the organic material completely, so that the possibility that the cobalt will be firmly complexed by residual organic matter, and not be available for determination, is avoided.

One intercomparison[1] showed that digestion with sulphuric acid and potassium perchlorate was the only satisfactory procedure out of a list including digestion with mixtures of nitric and sulphuric acids, nitric acid and potassium perchlorate, sulphuric acid and potassium permanganate, and sulphuric acid and hydrogen peroxide, but there are many more reports of investigations which indicate that some at least of these mixtures are satisfactory. There was no comment as to why sulphuric and perchloric acids were not used instead of the acid and salt mixture selected.

3. Nickel

Much of the interest in the determination of nickel in organic materials arises from its importance in petroleum products where it can be a serious catalyst poison. The main investigations into its recovery have therefore been concerned with the determination of traces of nickel in oil fractions.

For the analysis of organo-nickel compounds, or other organic materials containing much nickel, the comments made for cobalt largely apply again. Reduction to the metal is the almost unanimous choice of the textbooks, and many simple compounds can be determined colorimetrically or by titration, without further treatment. X-ray fluorescence procedures have been applied to the determination of nickel in petroleum products down to very low levels, as also have atomic absorption techniques.

[1] Sporek, K. F., *Anal. Chem.* **33**, 754 (1961).

Atomic absorption is proving particularly valuable for oil products as the samples are themselves readily inflammable, and can be sprayed directly, although they are more commonly diluted with a suitable solvent.

Investigations into the total oxidation of organic samples prior to the determination of nickel have largely been concerned with petroleum products, but as nickel is more likely to occur in volatile organic combination in oil than in most products, the methods which are satisfactory for these samples should be suitable for most others.

Most of the comparisons have been between straight-forward burning and ashing, and burning and ashing after the addition of sulphuric acid and the results have not always been consistent. Some investigations have found little difference between the methods,[1] but most workers have found significant differences for overhead fractions[2,3] with recoveries after sulphated ashing being up to 75% higher,[3] although similar differences were not observed with crudes and residues.[2] This latter view of the desirable effect of sulphated ashing is borne out by work with the nickel tetraphenyl porphyrin which is believed to be one of the forms in which nickel occurs in petroleum.[4] This revealed that recoveries were low after normal dry ashing but good when the oil was first sulphated.

A more recent technique for destroying the nickel porphyrin complex in oil is the treatment of the oil with benzene sulphonic acid. After this treatment the sample can be burned and ignited without any further addition of sulphuric acid.[5] The addition of one-tenth of its weight of sulphur to the oil before ignition has also been found to prevent losses of nickel.[6]

For samples other than petroleum products a variety of wet and dry methods have been used, although the interest here is much less. Again there does not seem to be much risk of loss, and the comments made on oxidation methods for the recovery of cobalt probably apply equally to the recovery of nickel.

[1] Barney, J. E., and Haight, G. P., *Anal. Chem.* **27**, 1285 (1955).
[2] Milner, O. I., Glass, J. R., Kirchner, J. P., and Yurick, A. N., *Anal. Chem.* **24**, 1728 (1952).
[3] Gamble, L. W., and Jones, W. H., *Anal. Chem.* **27**, 1456 (1955).
[4] Horeczy, J. T., Hill, B. N., Walters, A. E., Schutze, H. G., and Bonner, W. H., *Anal. Chem.* **27**, 1899 (1955).
[5] Shott, J. E., Garland, T. J., and Clark, R. O., *Anal. Chem.* **33**, 506 (1961).
[6] Agazzi, E. J., Burtner, D. C., Crittenden, D. J., and Patterson, D. R., *Anal. Chem.* **35**, 332 (1963).

One attempt to apply the oxygen flash technique to the oxidation of organo-nickel compounds was abandoned due to the formation of insoluble oxides.[1]

P. The Platinum Metals
Ruthenium, Rhodium, Palladium, Osmium Iridium, and Platinum

The six elements ruthenium, rhodium, palladium, osmium, iridium and platinum are all metals of high melting points, which are, in the main, readily liberated from their compounds. Ruthenium and osmium are the only two elements which form tetroxides, and these are very volatile. These facts largely govern the choice of methods which can be used to separate the elements from organic matrices, and indicate the need to consider ruthenium and osmium separately from the other four metals.

Rhodium, palladium, iridium and platinum should not present major difficulties in any of the main wet oxidation procedures, although dry ashing methods might be a little more troublesome. Very little has been published on the recovery of these metals, but any of the standard mixtures of sulphuric, nitric and perchloric acids, and hydrogen peroxide should prove reasonably suitable. This view is supported by work which measured the volatilization of many elements when heated with various mixtures of sulphuric, phosphoric, perchloric, hydrochloric and hydrobromic acids.[2] In no case was any loss of these four metals recorded.

In view of the ease with which the platinum metals are reduced to the metallic state, some problems can be expected if dry ashing methods are used. The most likely cause of trouble would be diffusion of the free metal into the structure of the ashing vessel and the magnitude of this problem will obviously depend on the ashing temperature, the nature of the ash, and the presence of any ashing aid. It is likely to occur with all the common types of vessel, platinum, porcelain and silica, although there is no evidence for the superiority, or otherwise, of any one of them. The use of a low ashing temperature, and a bulky ashing aid such as magnesium nitrate, would probably be sufficient to give reasonable recoveries.

Despite this possible drawback, ignition to the element and weighing is quoted in many textbooks as the method of choice for determining platinum in organic compounds. Of course, when the residue is to be

[1] Belcher, R., McDonald, A. M. G., and West, T. S., *Talanta*, **1**, 408 (1958).
[2] Hoffman, J. I., and Lundell, G. E. F., *J. Res. Nat. Bur. St.* **22**, 465 (1939).

weighed in the vessel in which the compound is ignited, it is unimportant if part of the metal has diffused into it.

One possible cause of difficulty that might be encountered in both wet and dry ashing procedures is the problem of redissolving the metal if once it has been produced by reduction. Rhodium and palladium should not present great difficulty, but platinum and iridium may require special procedures.

The recovery of osmium and ruthenium from organic matrices is a rather different situation, as the main risk is loss by volatilization. The volatile form in each case is the tetroxide, and high-temperature oxidation systems, such as are used in the usual wet digestions, are all likely to convert the elements to this form. Quantitative distillation of ruthenium and osmium from perchloric acid solution has been demonstrated[1] as have losses from solution containing hydrogen peroxide[2] and nitric acid.[3] One solution to the problem, is therefore to make use of the volatility and combine the wet oxidation step with a distillation to separate and recover the elements. This type of procedure has been claimed to be only 90 to 95% complete for osmium[4] and treatment of the osmium compound with sodium peroxide and sodium carbonate, and heating over a small flame for 1 hr was recommended. Incomplete recovery of microgram amounts of osmium on distillation from nitric acid have also been reported, although larger amounts could apparently be recovered satisfactorily.[5]

Normal dry ashing methods have been used with success, particularly for ruthenium, presumably because of the reducing conditions initially established. Investigation has concentrated on ruthenium because of the importance of the fission product, ruthenium 106, which tends to concentrate in certain types of seaweed forming a link in some of the food chains leading to man. An Atomic Energy Authority report[6] has shown that the loss of ruthenium on ashing such sea weeds at temperatures of 450 to 500°C was not more than 5%.

[1] Hoffman, J. I., and Lundell, G. E. F., *J. Res. Nat. Bur. St.* **22,** 465 (1939).
[2] Down, J. L., and Gorsuch, T. T., *Analyst,* **92,** 398 (1967).
[3] Sandell, E. B., *Ind. Eng. Chem. (Anal. Ed.),* **16,** 342 (1944).
[4] Dwyer, F. P., and Gibson, N. A., *Analyst,* **76,** 104 (1951).
[5] Bark, L. S., and Brandon, D. (and discussion), in Shallis, P. W., Ed., *Proc. S.A.C. Conf.,* Nottingham, July 19/23, 1965, p. 387.
[6] Riley, C. J., UKAEA Report PC 122 (W), 1962.

CHAPTER 9

SELECTED DECOMPOSITION PROCEDURES

There has been considerable controversy over the relative merits of wet and dry oxidation procedures, and it is not possible to state categorically that one is superior to the other.

Wet digestion has the advantage of being fairly rapid, requiring only simple apparatus and being less prone to suffer from volatilization and retention losses because of the liquid conditions and low temperatures involved. Its disadvantages are that the large amounts of reagents introduce blank problems, it will not conveniently handle large samples, and in general it requires considerable supervision.

By contrast dry ashing exhibits advantages and disadvantages that are almost the opposite of these. It will readily handle large samples, few or no reagents are added so that the blank problem is minimized, and little supervision is required. It suffers from being relatively time-consuming, from requiring more expensive equipment, and from a propensity to give rise to volatilization and retention losses because of the high temperatures and dry conditions involved.

In general it can be said that for the bulk of determinations these advantages and disadvantages can be weighed against each other merely on the basis of convenience, but for some applications there are overriding reasons for using one or the other. When volatilization will occur during dry ashing, as with mercury and selenium, then wet oxidation is almost essential. When very small traces are to be determined in large samples the same can well be said of dry ashing.

Dozens or even hundreds of different procedures have been described in the literature for the decomposition of organic samples, many of them differing only in unimportant details. The methods listed below are not claimed to be the only satisfactory ones, such a claim would be patently absurd. They are merely methods found to be satisfactory in fairly extended use.

A. Wet Digestion Methods

These methods are somewhat less dependent on the nature of the sample because the large amounts of added reagents tend to suppress the effects of variations in composition. They are therefore probably the first choice when handling a new type of sample, and a wet oxidation procedure can be devised for the recovery of virtually any element.

Open flasks are frequently used for these digestions, with Kjeldahl or conical flasks being particularly popular. However, the type of apparatus shown in Fig. 9.1 has many advantages in terms of flexibility, and its use can be recommended. The key feature is the three position tap A

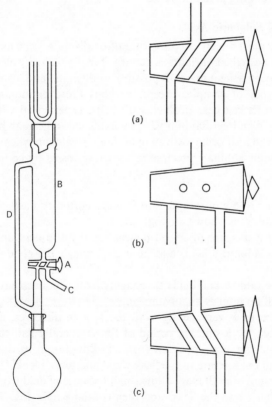

FIG. 9.1. Apparatus for controlled decomposition of organic material.

which is used to allow refluxing (position (a)) or distillation (position (b)) and removal of the distillate (position (c)). By its use, the oxidation potential of the system can be stabilized indefinitely by allowing the acid mixture to reflux, or it can be raised by allowing distillation to occur, and then again stabilized by running out the distillate and returning to the refluxing position. Fractions of distillate can be discarded, reserved for further treatment, or examined at leisure in the course of an investigation. In some cases it is convenient to use a two necked flask instead of the one shown, and to insert a thermometer into the flask. By this means, the lengths of the distillation stages can be controlled by the temperature attained by the reaction solution: this can be of particular value with mixtures containing perchloric acid.

1. Nitric and Sulphuric Acids

This is the most widely used mixture of oxidizing agents, and in one or other of the described variants is suitable for the recovery of virtually all elements except selenium, and possibly ruthenium and osmium, from nearly all types of sample, except, perhaps those containing covalent chlorine. In this last case problems might be experienced with elements such as arsenic and germanium where volatile chlorides may be formed.

Problems can also occur with samples high in calcium where the formation of a calcium sulphate precipitate can cause losses by coprecipitation.

Procedure I

1. Weigh up to 5 g of sample into a round bottomed flask, and connect to the apparatus as shown in Fig. 9.1.

2. Precool a mixture of 25 ml concentrated nitric acid and 10 ml of concentrated sulphuric acid, and pour it, through the condenser, on to the sample.

3. Mix the sample and acids thoroughly. If it is convenient a standing period of half an hour is an advantage.

4. Heat the flask very gently until the sample dissolves, and heat to boiling. Reflux for a suitable period of time, generally half to one hour.

5. Turn the tap *A* to position (b) and allow liquid to distil into the reservoir until the mixture in the flask chars and turns black.

6. By manipulating the tap, allow small volumes of acid from the reservoir to run back into the flask, and then continue refluxing. The charred residue should first clear and then darken again.

7. Repeat step 6 until the mixture in the flask no longer chars on heating to fumes.

8. Carefully allow all the solution in the reservoir to run back into the flask and reflux for 10 min.

9. Turn the tap to position (b) and heat to fumes of sulphuric acid. If the mixture darkens repeat steps 6 to 9. If it does not darken allow to cool and run the distillate out of the reservoir. This can be rejected or reserved for further consideration.

10. Add 15 ml of water to the residue in the flask, turn tap to position (b), and heat to fumes of sulphuric acid.

The elements to be determined remain in the residue in the flask, and can be treated further as appropriate.

This procedure is not entirely satisfactory with (i) very wet samples, (ii) those containing chloride ion, or (iii) those which react very vigorously with nitric acid. For (i) use procedure II, and for (ii) and (iii) procedure III or IV.

Procedure II

Proceed as in procedure I to step 5. At this point the distilled acid mixture in the reservoir may well be too weak to complete the oxidation successfully. Continue:

5. Run the acid mixture from the reservoir, and replace it with 25 ml of concentrated nitric acid.

6. Continue as in procedure I.

Procedure III

1. Weigh up to 5 g of sample into a round bottomed flask and connect to the apparatus as shown in Fig. 9.1.

2. Precool a mixture of 10 ml concentrated sulphuric acid and 20 ml distilled water and pour it through the condenser on to the sample.

3. Mix the sample and acids thoroughly.

4. Heat the mixture to boiling, and reflux for half an hour.

5. Turn the tap A to position (b) and allow liquid to distil into the reservoir until the sample chars. Allow to cool.

6. Turn the tap to position (c) and reject the distillate.

7. Return the tap to position (a) and carefully add 25 ml concentrated nitric acid by way of the condenser.

8. Heat the mixture to boiling and proceed as from step 5 procedure I.

Procedure IV

1. Weigh up to 5 g of sample into a round bottomed flask and connect to the apparatus as shown in Fig. 9.1.

2. Add 20 ml of water to the sample, mix thoroughly, and add 10 ml of concentrated nitric acid, 2 or 3 ml at a time, mixing well after each addition.

3. Warm the flask gently, and gradually increase to full reflux. Reflux for half to one hour.

4. Add 10 ml of concentrated sulphuric acid by way of the condenser.

5. Turn the tap A to position (b) and allow liquid to distil into the reservoir until the mixture chars. Allow to cool.

6. Turn the tap to position (c) and reject the distillate.

7. Return the tap to position (a) and carefully add 25 ml concentrated nitric acid through the condenser.

8. Heat the mixture to boiling and proceed as from step 5 procedure I.

Procedure V

When recovering mercury from organic material by procedure I it has been shown that an appreciable fraction of the mercury distils into the reservoir. To overcome this problem the Analytical Methods Committee of the Society for Analytical Chemistry has described a method which involves returning the distillate to the flask containing the residue, and carrying out the determination on the mixture. The apparatus used is similar to that shown in Fig. 9.1, but with the substitution of a two-necked flask, carrying a thermometer, for the single-necked flask shown. The procedure is as follows:

1. Weigh up to 10 g of sample into the flask, add 20 ml water, 5 ml concentrated sulphuric acid and 50 ml of nitric acid and assemble the apparatus.

2. Heat the flask gently at first, and then more strongly. With the tap A in position (b) collect the distillate until the temperature reaches 116°C.

3. Turn the tap to position (c) and run off, and reserve, the collected distillate.

4. Continue as for steps 5 to 7 in procedure I.

5. Allow to cool, run the distillate back into the flask, and add the distillate collected in step 3.

For very wet samples the amount of water added in step 1 can be reduced. The reason for removing the first distillate fraction when the temperature reaches 116°C in step 3 is that this is the point at which excess water has been removed and approximately concentrated nitric acid remains in the flask. If it should prove more convenient, this approach may be used instead of the total rejection of the diluted acid described in procedures II and IV.

2. Mixtures Containing Perchloric Acid

Perchloric acid can be used as a small final addition in a procedure based mainly on other oxidants, or as a major constituent present in the system from the start. In both cases the precautions described for handling perchloric acid must be observed. When the use of say nitric and sulphuric acids fails to yield a colourless solution in step 10 of procedure I the following addendum can be used.

11. Cool the flask. Mix 1 ml of 72% perchloric acid and 2 ml concentrated nitric acid, and add to the flask by way of the condenser. Reflux for 10 min.

12. Heat to fumes. Allow to cool.

If the solution is still coloured, it is very unlikely that this is due to unoxidized organic matter.

When perchloric acid forms a major constituent from the start of the digestion effective oxidations can be carried out with or without the presence of sulphuric acid. The presence of sulphuric acid is desirable wherever possible, but when precipitation of insoluble sulphates is likely it may be better to omit it. For the determination of selenium, mixtures containing perchloric acid are almost essential for maintaining oxidizing conditions, but osmium and ruthenium will distil quantitatively from these mixtures. The presence of perchloric acid has been shown to convert antimony to the pentavalent state in which it will react with rhodamine B, but this might possibly cause trouble with some materials containing covalent chlorine where the low boiling point of antimony pentachloride (79°C) and the generation of hydrochloric acid at fairly high temperatures might lead to difficulties. This type of loss has not been demonstrated, and quite probably will not occur due to rapid hydrolysis, but it is a point worth investigating.

Whether sulphuric acid is present or not, a common procedure can be used for perchloric acid–nitric acid mixtures.

Procedure VI

1. Weigh up to 5 g of sample into a round bottomed flask and connect to the apparatus as shown in Fig. 9.1.
2. Mix 25 ml concentrated nitric acid and 10 ml 72% perchloric acid. Add 5 ml concentrated sulphuric acid if appropriate.
3. Cool the mixture and add it to the sample by way of the condenser.
4. Mix the sample and acids thoroughly, and if convenient allow to stand for half an hour.
5. Warm gently, and gradually increase heating to full reflux. Heat for half to one hour.
6. Turn the tap to position (b) and allow the mixture to distil into the reservoir until the reaction becomes more vigorous.
7. Turn the tap to position (c), run the distillate out of the reservoir, return the tap to position (a) and reflux for half an hour.
8. Repeat steps 6 and 7 as often as necessary to achieve a smooth oxidation. This can generally be achieved in two or three steps, of which the final one is evaporation to fumes of perchloric acid.

If a two-necked flask and a thermometer are used, the distillation periods can be controlled by the temperature rise. Holding temperatures of 140, 180 and 200°C will usually be satisfactory, but these can be varied at will to suit the samples.

In all oxidations with perchloric acid the organic material should not be allowed to char. If this happens remove the source of heat and run some of the distillate back into the flask.

When using mixtures not containing sulphuric acid, care must be taken not to allow the flask to boil dry.

3. Sulphuric Acid and Hydrogen Peroxide

This is a vigorous procedure which has been shown to cause some problems with volatile elements. An effective procedure is described as procedure VII, but preliminary recovery trials are desirable with new combinations of sample and elements. Troubles have been found with selenium, and ruthenium, and, with germanium and arsenic in the presence of chlorine. Osmium and mercury will almost certainly also give low recoveries.

Procedure VII

1. Weigh up to 5 g of sample into a round bottomed flask and connect to the apparatus as shown in Fig. 9.1.

2. Add 40 ml water and 20 ml concentrated sulphuric acid to the sample through the condenser.

3. Turn the tap *A* to position (b) and heat until the sample is heavily charred.

4. Discard the distillate, cool slightly, and add 10 ml of 50% hydrogen peroxide to the sample in small amounts by way of the condenser.

5. Allow to reflux for a few minutes, then turn the tap to position (b) and heat until the sample chars.

6. By manipulating the tap *A* allow a few millilitres of distillate to run back into the flask. If the sample clears, return the tap to position (b) and again heat to charring.

7. Repeat step 6 until the sample no longer clears when a small quantity of the distillate is run into the flask.

8. Repeat steps 4 to 7 until the solution no longer darkens.

9. Heat to fumes of sulphuric acid, reject the distillate, cool, add 10 ml of distilled water and reflux for 5 min.

The amount of sample that can be handled with the procedure as described depends largely upon its oxygen content. 5 g of carbohydrate would cause little trouble, but more than 2 g of an involatile hydrocarbon would probably lead to the consumption of all the sulphuric acid. The optimum proportions of acid and peroxide are best found by experiment.

B. Dry Ashing Procedures

It is much more difficult to describe a dry oxidation which will meet the needs of a wide range of samples, as the samples themselves introduce greater variation than is the case with wet digestion. Factors of importance in the recovery of individual elements are mentioned in the sections on the elements, but to give a basis for the necessary variations the following procedure should be generally suitable.

1. Weigh 5 or 10 g of sample into a suitable dish or crucible: it is desirable for the sample to be thinly spread over a fairly large area. Silica is probably the most generally useful construction material for the ashing vessel.

2. Add the ashing aid if appropriate (see below).

3. Dry and thoroughly char the sample. This is best carried out with an infra red lamp, but heating on a hot plate or very careful heating with a burner can be used alternatively. The sample must not be allowed to ignite, and, if a burner is used, care must be taken to avoid local overheating.

4. Introduce the ashing vessel carefully into a furnace heated to about 450°C. Again, the material must not be allowed to ignite. Slightly higher temperatures—about 500°C—may be used in some cases, particularly when a bulky ashing aid such as magnesium nitrate is used.

5. Heat overnight, or for a similar period. If unoxidized organic matter remains moisten the residue with water, or diluted nitric acid $(1 + 2)$, evaporate to dryness or a water bath and return to the furnace for a further period.

6. When a suitable ash has been obtained, cool the basin, and moisten the contents with a little water. Very carefully add 10 ml of $1 + 1$ hydrochloric acid and evaporate to dryness on a water bath. Take up the residue in very dilute hydrochloric acid $(1 + 9)$ or other suitable solvent.

Ashing Aids

It is a matter of opinion as to the best point at which the ashing aid should be added to the sample, although it is desirable to carry out the addition at an early stage to avoid the risk of loss during preliminary heating. This is particularly the case with substances such as sulphuric acid which are intended to modify the sample, perhaps by removing sodium chloride as hydrochloric acid at low temperature.

It is most convenient to add the ashing aid as a large volume of diluted material so that all parts of the sample are thoroughly wetted. Typical concentrations and amounts would be 10 ml of 10% sulphuric acid solution or 10 ml of 7% $Mg(NO_3)_2.6H_2O$ solution, for 5 g of sample. When the addition has been made, the ashing vessel is heated gently to drive off the added water, and then the ashing proceeds as described above.

APPENDIX I. NUCLEAR DATA ON RADIOACTIVE TRACERS

Element	Nuclide	Half-life	β emission MeV	β emission %	γ emission MeV	γ emission %	Daughter nuclide	Daughter Half-life	Comments
Lithium									No suitable nuclide.
Sodium	Na 22	2·6 years	0·5	90	0·5–1·3	280			Cyclotron produced.
Sodium	Na 24	15 hr	1·4	100	1·4–2·8	200			Half-life inconveniently short.
Potassium	K 42	12·4hr	2·0–3·6	100	1·5	18			Half-life inconveniently short.
Rubidium	Rb 86	18·7 days	0·7–1·8	100	1·1	9			
Caesium	Cs 134	2·1 years	0·1–0·7	99	0·5–0·8	213			
Caesium	Cs 137	30 years	0·5–1·2	100	0·66	86	Ba 137m	2·6 min	
Beryllium	Be 7	53 days	E.C.		0·5	12			
Magnesium	Mg 28	21·4 hr	0·4–2·9	200	0·4–1·8	230	Al 28	2·3 min	Low specific activity and short half-life.
Calcium	Ca 45	165 days	0·25	100					Soft β emission only.
Strontium	Sr 85	65 days	E.C.		0·5	100			No γ emission.
Strontium	Sr 89	51 days	1·5	100					Daughter half-life rather long.
Strontium	Sr 90	28 years	0·5–2·3	200			Y 90	2·9 days	No γ emission.
Barium	Ba 133	7·2 years	E.C.		0·1–0·4	146			
Barium	Ba 140	12·8 days	0·5–2·2	200	0·1–2·5	250	La 140	40 hr	Relationship of parent and daughter half-lives unsuitable.
Copper	Cu 64	12·8 hr	0·6–0·7	57	0·5	38			Half-life short.
Silver	Ag 105	40 days	E.C.		0·3–0·6	96			Cyclotron produced.
Silver	Ag 110m	253 days	0·1–0·5	98	0·5–1·6	316	Ag 110	24 sec	
Gold	Au 195	192 days	E.C.		0·1	12			Cyclotron produced. γ energy low.
Gold	Au 198	2·7 days	1·0	99	0·4	96			Half-life rather short.
Zinc	Zn 65	245 days	E.C.		1·1	49			
Cadmium	Cd 109	470 days	E.C.		0·02	2			Cyclotron produced. Difficult to count.
Cadmium	Cd 115m	43 days	1·6	97	0·5–1·3	82			
Mercury	Hg 203	47 days	0·2	100	0·3				Low γ abundance.
Scandium	Sc 46	84 days	0·4	100	0·9–1·1	200			Cyclotron produced.
Yttrium	Y 88	107 days	E.C.		0·5–2·7	200			
Yttrium	Y 90	64 hr	2·3	100					Short half-life. No γ emission.
Yttrium	Y 91	59 days	1·5	100					No γ emission.
Aluminium									No suitable nuclide.
Gallium	Ga 67	78 hr	E.C.		0·1–0·4	98			
Indium	In 114m	50 days	2·0	99	0·2–0·7	25	In 114	72 sec	Cyclotron produced. Rather short half-life.
Thallium	Tl 204	3·8 years	0·8	98					No γ emission.
Titanium									No suitable nuclide.
Zirconium	Zr 89	78 hr	E.C.		0·5–0·9	150			Cyclotron produced.
Zirconium	Zr 95	65 days	0·4–0·9	100	0·7	98	Nb 95	35 days	The quoted emissions are for the parent only. The long-lived daughter must be allowed for in experiments.

Element	Nuclide	Half-life	β emission MeV	%	γ emission MeV	%	Daughter nuclide	Half-life	Comments
Hafnium	Hf 175	70 days	E.C.		0·3	86			For tracer studies a mixture of these two isotopes obtained by neutron irradiation of the natural material is quite suitable.
	Hf 181	45 days	0·4	96	0·5	81			
Germanium	Ge 68	280 days	1·9	86	0·5	174	Ga 68	68 min	The usable radiation is from the gallium daughter, so time for equilibration must be allowed before counting.
Tin	Sn 113	118 days	E.C.		0·4	67	In 113m	104 min	The usable radiation is from the daughter—see note above.
Lead	Pb 210	22 years	1·17	100			Bi 210 / Po 210	5 days / 138 days	The usable β radiation is from the daughter Bi 210. The grand-daughter polonium 210, is an α emitter and introduces a serious health hazard.
	Pb 212	10·6 hr	2·3	54	0·24	36	Bi 212 / Po 212	60 min / 3×10^{-7} sec	This tracer can be separated from aged thorium salts. The usable β radiation, and some γ radiation comes from the Bi 212 daughter, and equilibration is necessary before counting. The half-life is inconveniently short.
Vanadium	V 48	16 days	0·7	56	0·5-1·3	310			Cyclotron produced.
Niobium	Nb 95	35 days	0·16	100	1·0-1·3	100			
Tantalum	Ta 182	115 days	0·2-0·5	100	0·5-0·6	94			
Arsenic	As 74	18 days	0·7-1·5	62	0·6	135			Cyclotron produced.
	As 76	27 hr	2·4-3·0	87		57			Half-life rather short.
Antimony	Sb 122	2·7 days	0·7-2·0	97	0·6-2·1	66			Half-life rather short.
	Sb 124	60 days	0·6-2·3	87	0·4-0·6	185			
	Sb 125	2·7 years	0·1-0·6	97		77	Te 125m	58 days	More expensive than Sb 124. Daughter does not contribute to β or γ counts, so equilibration not usually required.
Bismuth	Bi 206	6·3 days	E.C.		0·2-1·7	385			Cyclotron produced.
	Bi 207	28 years	E.C.		0·6-1·1	174			Cyclotron produced.
Chromium	Cr 51	28 days	E.C.		0·3	8			Counting sensitivity low for most methods.
Molybdenum	Mo 99	67 hr	0·5-1·2	100	0·1-0·8	114	Tc 99m	6 hr	Daughter contributes largely to count so equilibration is required.
Tungsten	W 185	74 days	0·4	100	0·1-0·8	102			β emission rather low energy.
	W 187	24 hr	0·5-1·3	99	0·1-0·4	167			
Selenium	Se 75	121 days	E.C.						Half-life short.

Element	Isotope	Half-life	β (MeV)	β %	γ (MeV)	γ %	Daughter	Daughter half-life	Remarks
Tellurium	Te 132	78 hr	0·2-2·1	200	0·2-2·2	380	I 132	2·3 hr	Daughter contributes largely to count so equilibration is required.
Polonium	Po 210								α emitter. Requires special precautions and special counting methods.
Manganese	Mn 52	5·7 days	0·6	29	0·5-1·4	358			Half-life short for some requirements. Cyclotron produced.
	Mn 54	314 days	E.C.		0·8	100			Cyclotron produced.
Rhenium	Re 183	68 hr			0·1-0·3	40			A mixture of these isotopes is produced by neutron irradiation of natural rhenium. Both half-lives rather short.
	Re 186	90 hr	0·9-1·1	96	0·14	10			
	Re 188	17hr	2·0-2·1	98	0·16	10			
Iron	Fe 59	45 days	0·3-0·5	99	1·1-1·3	100			
Cobalt	Co 56	77 days	1·5	18	0·5-3·5	290			Cyclotron produced.
	Co 57	270 days	E.C.		0·1	98			Cyclotron produced.
	Co 58	71 days	0·5	15	0·5-0·8	130			Reactor produced, but available carrier free.
	Co 60	5·3 years	0·3	100	1·2-1·3	200			
Nickel	Ni 63	120 years	0·07	100					Only very soft β emission.
Ruthenium	Ru 103	40 days	0·2	89	0·5	88			β emission rather low energy.
	Ru 106	1 year	2·0-3·5	99	0·5-0·6	31	Rh 106	30 sec	Usable radiation is from Rh 106.
Rhodium	Rh 105	36 hours	0·6	90	0·3	10			Half-life short.
Palladium	Pd 103	17 days	E.C.	100	0·02		Ag 109m	40 sec	Only low energy X-rays.
Osmium	Os 191	15 days	0·14		0·13	25	Ir 191m	4·9 sec	γ- and X-rays come from the daughter nuclide.
Iridium	Ir 192	74 days	0·5-0·7	89	plus X-rays 0·30-0·47	175			
Platinum	Pt 197	20 hr	0·7	94	0·08	19			Short half-life and weak γ rays.

Notes on the table
1. β emission includes positrons.
2. γ emission includes X-rays.
3. The emissions quoted include radiation from the daughter nuclide when appropriate.
4. The percentage emissions include only the more abundant radiation.
5. The lanthanide and actinide elements are not included in the table.

INDEX

Italic figures indicate principal references

Actinium 88-9
Activation analysis 3
Adsorption 14, 15, 64
Aluminium 5, 14, *84 et seq.*
Animal tissues 3, 57, 70, 85, 95, 105, 114
Antimony 14, 34, 41, 100, *109-11*, 141
Arsenic 8, 14, 27, 33, 41, *100-9*, 138, 142
Ashing aids *37 et seq.*, 64, 74, 78, 89, 96, 97, 98, 99, 104, 105, 108, 110, 112, 115, 116, 118, 125, 126, 127, 130, 131, 134, 144
Atomic absorption 6, 61, 69, 73, 105, 122, 132, 133

Barium 14, *67 et seq.*
Benzene sulphonic acid 113, 133
Beryllium *67 et seq.*
Bismuth 33, 100, *111-12*
Blood 8, 41, 61, 66, 69, 70, 73, 81, 96, 105, 110, 114, 122, 125, 130, 131
Boric acid 38, 99
Bromine 7, 8, 67, 94

Cadmium 33, 35, 41, *73 et seq.*
Caesium 17, 41, 52, 53, *55 et seq.*
Calcium 14, 18, 36, *67 et seq.*
Carius oxidation 73, 80, 94, 102, 118
Cerium 41
Chlorine 27, 37, 41, 90, 91, 107-8
Chromium 41, *114-16*
Coal 89-90
Cobalt 17, 41, 121, *129-32*
Cocoa 66, 78, 91, 95, 96, 100, 105, 106, 110, 112, 124, 130
Colorimetry 5, 122, 130, 132
Copper 7, 10, 13, 14, 36, 41, *60 et seq.*

Electron spin resonance 4
Excited oxygen *39*, 65, 67, 84, 90, 104, 119, 126

Feeding stuffs 61, 71, 117, 123, 131
Flame photometry 6, 60, 69

Gallium 33, *84*
Gamma counting 50
Gamma spectrometry 46, 76
Gas chromatography 5, 94
Geiger counting 48
Germanium 27, *89 et seq.*, 138, 142
Glassware 12, *14-17*, 59, 85, 96
Gold 13, 15, 41, *60 et seq.*

Hafnium 99-100
Hydrobromic acid 93
Hydrochloric acid 6, 7, 15, 63, 67, 76, 79, 85, 86, 94, 99, 103, 122, 128, 130, 141, 144
Hydrogen peroxide 19, 20, *24*, 25, 61, 66, 73, 74, 78, 80, 83, 86, 88, 90, 93, 94, 95, 100, 103, 106, 107, 108, 109, 111, 112, 114, 115, 117, 121, 123, 124, 132, 134, 135, 142, 143

Indium *84 et seq.*
Iridium 13, 41, *134-5*
Iron 1, 7, 10, 13, 34, 41, *121-9*

Lead 4, 5, 7, 13, 18, 21, 33, 34, 35, 38, 89, *93 et seq.*
Lithium *55 et seq.*

Magnesium 36, *67 et seq.*
Magnesium nitrate 38, 63, 64, 96, 97, 104-5, 109, 110, 130, 131
Manganese 10, 14, 16, 41, *116-17*
Mass spectrometry 5

Mercury 6, 8, 16, 31, 33, 34, 39, 41, *73 et seq.*, 136, 140, 142
Milk 23, 61, 95, 99
Molybdenum 114–16

Nickel 15, 33, 121, *132–4*
Niobium 112–14
Nitric acid 7, 17, 19, 20, *21*, 22, 23, 25, 27, 37, 41, *42*, 61, 62, 63, 65, 70, 72, 74, 78, 80, 82, 88, 90, 92, 93, 94, 95, 100, 102, 103, 106, 107, 108, 109, 110, 111, 114, 115, 116, 117, 118, 120, 121, 123, 124, 131, 132, 134, 135, 138, 140, 141, 142

Osmium 26, *134–5*, 138, 141, 142
Oxygen flask 31, *32 et seq.*, 71, 73, 80, 84, 94, 101–2, 118, 119, 134
Ozone 42

Palladium 15, *134–5*
Parr bomb 40, 109
Perchloric acid 5, 7, 15, 19, *22 et seq.*, 61, 62, 65, 72, 74, 78, 83, 88, 92, 93, 94, 95, 102, 106, 107, 108, 110, 111, 114, 115, 116, 117, 118, 120, 121, 123, 124, 132, 134, 141, 142
Petroleum products 5, 6, 7, 29, 60, 61, 69, 71, 73, 94, 103, 112, 113, 122, 127, 132, 133
Plant materials 3, 70, 80, 83, 85, 90, 95, 105, 114, 124
Platinum 15, *134–5*
Platinum apparatus 13, 36, 64, 69, 74, 88, 129, 132, 134
Plutonium 16, *87–9*
Polonium 117, 121
Polyethylene 41, 78, 91, 106, 107, 121
Polyethylene apparatus 17, 85
Polyvinyl chloride 35, 78, 91, 99, 106, 107, 108, 115, 121
Porcelain apparatus 13, 35, 62, 65, 88, 128, 131, 132, 134
Potassium 14, *55 et seq.*
Potassium nitrate 40, 100, 101
Potassium perchlorate 40, 132
Potassium permanganate 8, 81, 102, 103, 109, 132

Potassium sulphate 100
Precipitates 18, 26, 45, 64, 95
Procedures 136–44
Protactinium 88–9

Radium 67
Rare earths 87 *et seq.*
Retention losses 35–7, 38, 45, 59, 62, 67, 75, 87, 96, 97, 98, 99, 110, 113, 116, 117, 128, 131, 132, 134, 136
Rhenium 41, *116*
Rhodium 134–5
Rubidium 55 *et seq.*
Ruthenium 15, 26, 41, *134–5*, 138, 141, 142

Scandium 87 *et seq.*
Scintillation counting 51
Selenium 39, 41, *117–21*, 136, 138, 141, 142
Silica apparatus 12, 13, 35, 36, 63, 69, 74, 88, 110, 128, 132, 134, 143
Silver 3, 13, 16, 17, 41, *60 et seq.*
Sodium 14, 41, *55 et seq.*
Sodium chloride 34, 36, 55, 58, 75, 76, 77, 91, 98, 106, 110, 111, 114, 116, 127, 128, 144
Sodium nitrate 40
Sodium peroxide 40, 102, 109, 118, 134
Sodium persulphate 80
Sodium sulphate 102
Strontium 41, *67*
Sulphuric acid 8, 15, *19 et seq.*, 21, 24, 25, 27, 38, 41, 61, 62, 64, 65, 66, 70, 72, 73, 74, 78, 81, 82, 83, 86, 87, 88, 90, 92, 93, 94, 95, 99, 100, 102, 103, 106, 107, 108, 109, 110, 111, 112, 114, 115, 116, 117, 118, 120, 121, 123, 124, 126, 127, 131, 132, 133, 134, 138, 139, 141, 142, 143, 144

Tantalum 112–14
Technetium 116
Tellurium 117–21
Thallium 41, *84 et seq.*
Thorium 88–9
Tin 89, *92–3*
Titanium 15, *99–100*

Titration 5, 70, 122, 130, 132
Trichloracetic acid 7, 122
Tungsten 114

Uranium 88–9
Urine 5, 6, 8, 69, 70, 73, 81, 88, 89, 114

Vanadium 5, 15, *112–14*
Vegetable material 3, 8, 70, 83, 85, 90, 95, 105, 114, 124

Volatilization losses *26*, 31, 33, *34–5*, 37, 40, 44, 45, 55, 59, 75, 78, 86, 87, 92, 96, 97, 98, 99, 106, 109, 110, 113, 115, 116, 119, 120, 127, 135, 136

X-ray absorption 4
X-ray flourescence 3, 70, 73, 112, 122, 130, 132

Yttrium *87 et seq.*

Zinc 5, 10, 14, 33, 34, 41, *73 et seq.*
Zirconium 99–100

OTHER TITLES IN THE SERIES IN
ANALYTICAL CHEMISTRY

Vol. 1. WEISZ—Microanalysis by the Ring Oven Technique. First edition (see Vol. 37).
Vol. 2. CROUTHAMEL—Applied Gamma-ray Spectrometry.
Vol. 3. VICKERY—The Analytical Chemistry of the Rare Earths.
Vol. 4. HEADRIDGE—Photometric Titrations.
Vol. 5. BUSEV—The Analytical Chemistry of Indium.
Vol. 6. ELWELL and GIDLEY—Atomic Absorption Spectrophotometry.
Vol. 7. ERDEY—Gravimetric Analysis Parts I-III.
Vol. 8. CRICHFIELD—Organic Functional Group Analysis.
Vol. 9. MOSES—Analytical Chemistry of the Actinide Elements.
Vol. 10. RYABCHIKOV and GOL'BRAIKH—The Analytical Chemistry of Thorium.
Vol. 11. CALI—Trace Analysis for Semiconductor Materials.
Vol. 12. ZUMAN—Organic Polarographic Analysis.
Vol. 13. RECHNITZ—Controlled-Potential Analysis.
Vol. 14. MILNER—Analysis of Petroleum for Trace Elements.
Vol. 15. ALIMARIN and PETRIKOVA—Inorganic Ultramicroanalysis.
Vol. 16. MOSHIER—Analytical Chemistry of Niobium and Tantalum.
Vol. 17. JEFFERY and KIPPING—Gas Analysis by Gas Chromatography.
Vol. 18. NIELSEN—Kinetics of Precipitation.
Vol. 19. CALEY—Analysis of Ancient Metals.
Vol. 20. MOSES—Nuclear Techniques in Analytical Chemistry.
Vol. 21. PUNGOR—Oscillometry and Conductometry.
Vol. 22. J. ZYKA—Newer Redox Titrants.
Vol. 23. MOSHIER and SIEVERS—Gas Chromatography of Metal Chelates.
Vol. 24. BEAMISH—The Analytical Chemistry of the Noble Metals.
Vol. 25. YATSIMIRSKII—Kinetic Methods of Analysis.
Vol. 26. SZABADVARY—History of Analytical Chemistry.
Vol. 27. YOUNG—The Analytical Chemistry of Cobalt.
Vol. 28. LEWIS, OTT and SINE—The Analysis of Nickel.
Vol. 29. BRAUN and TOLGYESSY—Radiometric Titrations.
Vol. 30. RUZICKA and STARY—Substoichiometry in Radiochemical Analysis.
Vol. 31. CROMPTON—The Analysis of Organoaluminium and Organozinc Compounds.
Vol. 32. SCHILT—Analytical Applications of 1,10-Phenanthroline and Related Compounds.
Vol. 33. BARK and BARK—Thermometric Titrimetry.
Vol. 34. GUILBAULT—Enzymatic Methods of Analysis.
Vol. 35. WAINERDI—Analytical Chemistry in Space.
Vol. 36. JEFFERY—Chemical Methods of Rock Analysis.
Vol. 37. WEIZ—Microanalysis by the Ring Oven Technique. (Second enlarged and revised edition).
Vol. 38. RIEMAN and WALTON—Ion Exchange in Analytical Chemistry.